On Formally Undecidable Propositions Of Principia Mathematica And Related Systems

KURT GÖDEL

Translated by
B. MELTZER

Introduction by
R. B. BRAITHWAITE

DOVER PUBLICATIONS, INC.
New York

This Dover edition, first published in 1992, is an unabridged and unaltered republication of the work first published by Basic Books, Inc., New York, in 1962.

Library of Congress Cataloging-in-Publication Data

Gödel, Kurt.
    [Über formal unentscheidbare Sätze der Principia Mathematica und verwandter Systeme I. English]
    On formally undecidable propositions of Principia mathematica and related systems / Kurt Gödel ; translated by B. Meltzer ; introduction by R.B. Braithwaite.
        p.   cm.
    Translation of a paper entitled Über formal unentscheidbare Sätze der Principia Mathematica und verwandter Systeme I, published 1931 in the Monatshefte für Mathematik und Physik, v. 38, p. 173–198.
    Reprint. Originally published: New York : Basic Books, c1962.
    ISBN-13: 978-0-486-66980-9 (pbk.)
    ISBN-10: 0-486-66980-7 (pbk.)
    1. Gödel's theorem.   I. Title.
QA248.G573   1992
511.3—dc20
                                            91-45947
                                            CIP

Manufactured in the United States by LSC Communications
66980717   2019
www.doverpublications.com

TO

CHRISTOPHER FERNAU

in gratitude

PREFACE

Kurt Gödel's astonishing discovery and proof, published in 1931, that even in elementary parts of arithmetic there exist propositions which cannot be proved or disproved within the system, is one of the most important contributions to logic since Aristotle. Any formal logical system which disposes of sufficient means to compass the addition and multiplication of positive integers and zero is subject to this limitation, so that one must consider this kind of incompleteness an inherent characteristic of formal mathematics as a whole, which was before this customarily considered *the* unequivocal intellectual discipline *par excellence*.

No English translation of Gödel's paper, which occupied twenty-five pages of the *Monatshefte für Mathematik und Physik*, has been generally available, and even the original German text is not everywhere easily accessible. The argument, which used a notation adapted from that of Whitehead and Russell's *Principia Mathematica*, is a closely reasoned one and the present translation—besides being a long overdue act of piety—should make it more easily intelligible and much more widely read. In the former respect the reader will be greatly aided by the Introduction contributed by the Knightbridge Professor of Moral Philosophy in the University of Cambridge; for this is an excellent work of scholarship in its own right, not only pointing out the significance of Gödel's work, but illuminating it by a paraphrase of the major part of the whole great argument.

I proposed publishing a translation after a discussion meeting on "Gödel's Theorem and its bearing on the philosophy of science", held in 1959 by the Edinburgh Philosophy of Science Group. I wish to thank this society for providing the stimulus, the publishers for their ready co-operation on the proposal, and Professor Braithwaite not only for the Introduction but also for meticulous assistance in translation and proof-reading of a typographically intricate text. It may be noted here that the pagination of the original article is shown in the margins of the translation, while the footnotes retain their original numbers.

B. MELTZER

University of Edinburgh
January, 1962

INTRODUCTION

by

R. B. BRAITHWAITE

Every system of arithmetic contains arithmetical propositions, by which is meant propositions concerned solely with relations between whole numbers, which can neither be proved nor be disproved within the system. This epoch-making discovery by Kurt Gödel, a young Austrian mathematician, was announced by him to the Vienna Academy of Sciences in 1930 and was published, with a detailed proof, in a paper in the *Monatshefte für Mathematik und Physik* Volume 38 pp. 173-198 (Leipzig: 1931). This paper, entitled *"Über formal unentscheidbare Sätze der Principia Mathematica und verwandter Systeme I"* ("On formally undecidable propositions of *Principia Mathematica* and related systems I"), is translated in this book. Gödel intended to write a second part to the paper but this has never been published.

Gödel's Theorem, as a simple corollary of Proposition VI (p. 57) is frequently called, proves that there are arithmetical propositions which are undecidable (i.e. neither provable nor disprovable) within their arithmetical system, and the proof proceeds by actually specifying such a proposition, namely the proposition *g* expressed by the formula to which "17 Gen *r*" refers (p. 58). *g* is an arithmetical proposition; but the proposition that *g* is undecidable within the system is not an arithmetical proposition, since it is concerned with provability within an arithmetical system, and this is a meta-arithmetical and not an arithmetical notion. Gödel's Theorem is thus a result which belongs not to mathematics

1

but to metamathematics, the name given by Hilbert to the study of rigorous proof in mathematics and symbolic logic.

METAMATHEMATICS. Gödel's paper presupposes some knowledge of the state of metamathematics in 1930, which therefore I shall briefly explain. Following on the work of Frege and Peano, Whitehead and Russell's *Principia Mathematica* (1910-13) had exhibited the fundamental parts of mathematics, including arithmetic, as a *deductive system* starting from a limited number of axioms, in which each theorem is shown to follow logically from the axioms and theorems which precede it according to a limited number of rules of inference. And other mathematicians had constructed other deductive systems which included arithmetic (see p. 37, n. 3). In order to show that in a deductive system every theorem follows from the axioms according to the rules of inference it is necessary to consider the *formulae* which are used to express the axioms and theorems of the system, and to represent the rules of inference by rules (Gödel calls them "mechanical" rules, p. 37) according to which from one or more formulae another formula may be obtained by a manipulation of symbols. Such a representation of a deductive system will consist of a sequence of formulae (a *calculus*) in which the initial formulae express the axioms of the deductive system and each of the other formulae, which express the theorems, are obtained from the initial formulae by a chain of symbolic manipulations. The chain of symbolic manipulations in the calculus corresponds to and represents the chain of deductions in the deductive system.

But this correspondence between calculus and deductive system may be viewed in reverse, and by looking at it the other way round Hilbert originated metamathematics. Here a calculus is constructed, independently of any interpretation of it, as a sequence of formulae which starts with a few initial formulae and in which every other formula is obtained from preceding formulae by symbolic manipulations. The calculus can then be interpreted as representing a deductive system if the initial formulae can be interpreted as expressing the axioms of the system and if the rules of symbolic manipulation can be interpreted as representing the logical rules of inference of the system. If this can be done, a proof that a formula (other than one of the initial formulae) occurs in the sequence of formulae of the calculus yields a proof that the proposition which is the interpretation of this formula is a theorem of the deductive system, i.e. can be deduced from the axioms of the system by the system's rules of inference. Metamathematicians in the 1920's established many important results about deductive systems by converting proofs of what formulae can be obtained by symbolic manipulations within a calculus into proofs of what theorems can be proved within a deductive system which could be represented by the calculus. Frequently consideration of symbolic manipulations provided a "decision procedure" by which whole classes of theorems could actually be proved. Thus Presburger in 1930 published a decision procedure applicable to every proposition of a mutilated system of arithmetic which uses the operation of addition but not that of multiplication; he proved that every one of its propositions is decidable, i.e. either provable or disprovable, within this system.

Gödel's paper established the opposite of this for an arithmetical system which uses multiplication as well as addition—"the theory of ordinary whole numbers" (p. 38). And this is the piece of mathematics which is oldest in the history of civilization and which is of such practical importance that we make all our children learn a great deal of it at an early age. Gödel was the first to prove any unprovability theorem for arithmetic, and his way of proof was subtler and deeper than the metamathematical methods previously employed. Either of these facts would have ranked this paper high in the development of metamathematics. But it was the fact that it was a proposition of whole-number arithmetic which he showed to be undecidable that created such a scandal.

GÖDEL'S 'FORMAL SYSTEM' P. In order rigorously to prove the undecidability of some arithmetical propositions it is necessary to be precise about the exact deductive system of arithmetic which is being considered. As is indicated in the title of his paper, Gödel takes for his arithmetical deductive system that part of the system of *Principia Mathematica* required to establish the theorems of whole-number arithmetic. Since his proof is metamathematical he is concerned with a calculus representing his arithmetical system: what he proves in Proposition VI (p. 57) is a result about the calculus and not about what the calculus represents, for what it directly establishes is that neither of two particular formulae—the first referred to by "17 Gen r", the second by "Neg (17 Gen r)" (p. 59)—can be obtained from the initial formulae of the calculus by the rules of symbolic manipulation of the calculus. If the calculus is interpreted (as it can be interpreted) so that it represents the arithmetical part of the *Principia Mathematica* deductive system, with the second formula expressing the contradictory of the arithmetical proposition expressed by the first formula, then the theorem about the deductive system which corresponds to the calculus-theorem states that the proposition g to which "17 Gen r" refers is such that neither it nor its contradictory is provable within the system. Hence within the system g is neither provable nor disprovable. An unprovability theorem for the arithmetical deductive

system which Gödel is considering is a simple corollary of
Proposition VI about his calculus. Thus the paper is concerned with what formulae can (or, rather, cannot) be
obtained within a particular calculus, although of course
the calculus would have little general interest if it could not
be interpreted as representing a deductive system of whole-
number arithmetic.

Gödel's attention solely to his calculus will explain
some features of his terminology which may puzzle philosophical logicians. He transfers many epithets which are
applied more naturally to deductive systems than to calculi,
using them to refer to features of his *formal system* (his term
for what I have called his calculus). He employs *formula* in
the way in which I have used it so that a formula is a "finite
series of basic signs", but he goes on to say that "it is easy to
state precisely just *which* series of basic signs are meaningful formulae and which are not" (p. 38). "Meaningful" is a
misnomer, since it is the formal system that is being considered and not an interpretation of it. When he specifies
on p. 43 precisely which series of basic signs are to be
well-formed formulae (to use the modern term)—Gödel calls
them formulae without a qualifying adjective—he makes
no reference to meaning. A formula for him is a series of
signs which either is an elementary formula (a concatenation
of signs of specified sorts) or is built up out of elementary
formulae together with some or all of three specified signs
by the use, or repeated use, of three specified rules of construction. When Gödel speaks, in connexion with a formal
system, of 'rules of inference', he is referring to the rules
according to which one formula can be obtained from other
formulae within the formal system. In his system he uses
two 'rules of inference', which he specifies by giving one
condition for a formula being an 'immediate consequence'
of two formulae and one condition for its being an 'im-

mediate consequence' of one formula (p. 45). A 'proof-schema', for him, is a series of formulae in which each formula (except the initial formulae, which he calls 'axioms') is an 'immediate consequence' of one or of two of the formulae preceding it in the series. A 'proof-schema' is a 'proof' of the last formula in it; and a formula is 'provable' if there is a 'proof' of it. Gödel gives his precise definition of the class of 'provable' formulae in language familiar to mathematicians as "the smallest class of formulae which contains the axioms and is closed with respect to the relation 'immediate consequence of'" (p. 45), i.e. the smallest class which contains the axioms and which contains the 'immediate consequence' of every formula, and of every pair of formulae, contained in the class. For the benefit of philosophical logicians I shall continue the practice followed in this paragraph of putting single quotation marks round terms which without quotation marks refer to features of deductive systems, when I am using them, in Gödel's manner, with reference to a formal system, i.e. to a calculus.

Gödel gives an "exact description" of his formal system P on pp. 42ff. by specifying (1) its basic signs, (2) its formulae (i.e. its well-formed formulae), (3) its 'axioms' (initial formulae), (4) the relation of being an 'immediate consequence' of. He says that P is "essentially the system obtained by superimposing on the Peano axioms [for whole-number arithmetic] the logic of PM [*Principia Mathematica*]" (p. 41). Since the Peano axioms are 'provable' (and indeed 'proved') in the calculus of PM, Gödel's system P is virtually that part of the calculus of PM required to lead up to whole-number arithmetic: as Gödel says, "the addition of the Peano axioms, like all the other changes made in the system PM, serves only to simplify the proof and can in principle be dispensed with" (p. 41, n. 16). Gödel states his rules of symbolic construction and manipulation more precisely than do Whitehead and Russell. His only noteworthy divergence from them is that, instead of employing a limited number of 'axioms', he follows the example of von Neumann in using, besides three of Peano's 'axioms', eight 'axiom-schemata' each covering an unlimited number of cases (p. 44): by doing this he is able to manage with only two 'rules of inference' (see p. 45, n. 24). Gödel specifies the formal system P in the way he does in order to simplify his proof of the undecidability of some of the formulae of P. Since, as he explains, this undecidability is not due to "the special nature of the systems set up, but holds for a very extensive class of formal systems" (p. 38), the exact form he has chosen for P is of no intrinsic importance. What is essential is that P should be an appropriate subject for the exhibition of a method of metamathematical proof which Gödel invented, a method so powerful that it can establish an 'unprovability' result for every formal system capable of representing arithmetic.

THE METHOD OF "ARITHMETIZATION". Gödel's novel metamathematical method is that of attaching numbers to the signs, to the series of signs (formulae) and to the series of series of signs ('proof-schemata') which occur in his formal system. Just as Descartes invented co-ordinate geometry by assigning number-pairs to the points of plane Euclidean geometry, so Gödel invented what might be called *co-ordinate metamathematics* by assigning numbers to the basic signs, series of basic signs, series of series of basic signs (all of which I shall for convenience lump together under the generic term *string*) which form an essential part of the subject-matter of metamathematics. Descartes proved geometrical theorems about points by proving algebraic theorems about numbers; Gödel established meta-mathematical results about the strings of his formal system by considering numbers co-ordinated with the strings. The difference between the co-ordinate systems of co-ordinate geometry and of co-ordinate metamathematics is that the former uses number-pairs for two-dimensional geometry, number-triads for three-dimensional geometry, and so on, and the numbers used are not confined to integers, whereas co-ordinate metamathematics is one-dimensional, using only single numbers, and these (in Gödel's paper) are restricted to being "natural numbers", i.e. 0, 1, 2, 3, etc.

Gödel explains what is now called his "arithmetization" method on p. 45. What he does is to provide a co-ordinating rule according to which a different number (which I shall call a *Gödel number*) is assigned to each string in his formal system. The rule also works in reverse: of every number 0, 1, 2, 3, etc. the rule determines whether the number is the Gödel number of a basic sign, or of a series of basic signs, or of a series of series of basic signs, or is not a Gödel number at all (i.e. there is no string of which it is the Gödel number); and if the number is a Gödel number, the rule specifies uniquely which string it is of which it is the Gödel number. In his account Gödel speaks of his rule as establishing a "one-to-one correspondence". Not all numbers are Gödel numbers: the one-to-one correspondence established by the rule is between the members of a specific sub-class of the class of natural numbers, namely those which are Gödel numbers, and the members of the class of strings, which class is the union (logical sum) of three exclusive classes—the class of basic signs, the class of series of these signs, the class of series of series of these signs. The Gödel number of a series of series of signs is not explicitly mentioned in Gödel's account of his method of arithmetization, but he uses the notion in the definitions (from 22 onwards, pp. 52ff.) which form an essential preliminary to the proof of his Theorem. This Gödel number is the number constructed out of the Gödel numbers of the elements of the series of series of signs in exactly the same way as the Gödel number of a series of signs is constructed out of the Gödel numbers of the signs which are the elements of this series of signs (p. 45). Thus the Gödel number of a series of k elements, whether these elements are signs or are series of signs, is constructed out of the elements' Gödel numbers $n_1, n_2, \ldots n_k$ as the number $2^{n_1} . 3^{n_2} \ldots p_k^{n_k}$, a product whose prime factors are the first k prime numbers (1 not being counted as a prime number) with the 1st, 2nd, $\ldots k$-th prime number occurring respectively $n_1, n_2, \ldots n_k$ times in the product. The one-to-one correspondence between Gödel numbers defined in this way and the strings of which they are Gödel numbers is a consequence of the "fundamental theorem of arithmetic", namely that every natural number greater than 1 which is not itself a prime has a unique resolution into prime factors.

Gödel's rule of arithmetization ensures that to every class of strings there corresponds a unique class of Gödel numbers, and *vice versa*. And that to any relation R between strings there corresponds a unique relation R' between Gödel numbers, and *vice versa*: i.e. the n-adic relation R' holds between n Gödel numbers if and only if the n-adic relation R holds between the n strings. For example, the metamathematical statement that the series s of formulae is a 'proof' of the formula f is true if and only if a certain arithmetical relation holds between the Gödel numbers of s and of f which corresponds to the relation: being a 'proof' of. Gödel uses the same language to refer to the arithmetical properties of, and relation between, Gödel numbers as he uses to refer to the corresponding properties of, and relations between, strings (see p. 39, n. 9), printing the terms in italics when they refer to arithmetical concepts applicable to Gödel numbers (p. 46). In a sequence of

definitions 6-46 (pp. 50-55) he defines, step by step, a
sequence of arithmetical concepts which correspond, according to his rule of arithmetization, to the metamathematical
concepts expressed by the same words. [Definitions 1-5
define the ancillary arithmetical concepts (being the n-th
prime number, etc.) used in his method of arithmetization.]
As examples, definition 8 (p. 50) defines the arithmetical
operation $*$ upon two numbers x and y in such a way that
the number $x * y$ which is the result of performing this
operation is the Gödel number of the string obtained by
taking the string whose Gödel number is x and placing the
string whose Gödel number is y immediately after it. And
definition 45 (p. 55) defines the arithmetical relation B
between x and y so that the proposition $x \, B \, y$ is the same
as the conjunction of the proposition that x is the Gödel
number of a series of series of signs forming a 'proof-
schema' with the proposition that the series of signs whose
Gödel number is y is the last series of signs in this 'proof-
schema', i.e. this 'proof-schema' is a 'proof' of the last
formula in it. For the sake of clarity I shall not follow
Gödel's abbreviating practice of using italicized words and
phrases to refer to arithmetical concepts applicable to
Gödel numbers, and shall use italics only in the ordinary
way for emphasis. For example, while Gödel paraphrases
the $x \, B \, y$ of definition 45 (p. 55) as: x is a *proof* of the
formula y, I paraphrase it as: x is the Gödel number of a
'proof' of the formula whose Gödel number is y.

The interpreted symbolism used in these definitions, as in
all Gödel's metamathematical statements (see p. 47, n. 29),
is that of Hilbert and Ackermann's *Grundzüge der theoretischen Logik* (1928: English translation, 1950). The only
deviations from the symbolism of *Principia Mathematica*
are: "$\bar{p}$" to stand for Not p, "$p \, \& \, q$" for Both p and q,
"$p \rightarrow q$" for Not both p and Not q (the "$p \supset q$" of PM),

"$p \sim q$" for Either both p and q or both Not p and Not q (the "$p \equiv q$" of PM), and "(Ex)" as the existential quantifier in place of the "$(\exists x)$" of PM. Gödel uses "$\equiv$" as an abbreviation for "means the same as" in his definitions.

Except for these purely logical concepts, all the concepts involved in Gödel's definitions 1-46 (pp. 49-55), and also those in (5), (6), (6.1), (8.1) of pp. 57f., are arithmetical concepts (properties, relations, operations) which apply to natural numbers, i.e. the substitution values for the variables "x", "y", "z", "n", etc. occurring in the definitions are "0", "1", "2", . . . And the logical concepts are restricted so that they apply to only a finite number of entities. Whenever a universal or existential quantifier occurs in any of the definitions 1-45, a clause is inserted within the quantification which ensures that the quantification is only over a finite number of values. For example, the first definition (p. 49) defines "x is divisible by y" as There is a *z less than or equal to x* which is such that $x = y.z$, the phrase which I have italicized being inserted so as to restrict the quantification to numbers not greater than x. This makes the definiens equivalent to: $x = y.0$ or $x = y.1$ or . . . or $x = y.x$, a truth-function of a finite number $(x+1)$ of equalities. This restriction upon the quantifiers (except in definition 46) secures that all the arithmetical concepts employed (except Bew) are *recursive* in a sense of this word which Gödel defines and discusses in an excursus from his main argument (pp. 46-49).

RECURSIVENESS. The notion of recursiveness has played a central part in metamathematics since Gödel's work on it, but little more will be said about it here than is necessary for an understanding of Gödel's proof of his Theorem.

The method of *recursive definition* is an extension of the method of definition by "mathematical induction" by which the natural numbers are, step by step, defined. Starting with 0, 1 is defined as the immediate successor of 0, 2 as the immediate successor of 1, and so on. A recursive definition (a "primitive recursive definition", as it is now called) is the specification of each number in a sequence of numbers by means of a specification of the first number and of a rule which specifies the $(k+1)$-th number in terms of the k-th number and of k itself. [This is a paraphrase of Gödel's definition of a *recursively defined* arithmetical function, where this function is of only one numerical variable: see (2) of p. 46.] An arithmetical function is *recursive* if it is the last term in a finite sequence of functions in which each function is recursively defined by a rule involving two functions preceding it in the sequence (or is the successor function or a constant or obtained by substitution from a preceding function); and the recursiveness of other arithmetical concepts is defined by means of the notion of recursive function. The essential feature of a recursive concept—a dyadic relation R, for example—is that whether or not R holds between m and n, i.e. whether $R(m, n)$ is true or false, can be decided by a step-by-step procedure working upwards from $R(0, 0)$ with the use of a limited number of recursive definitions.

The importance of recursiveness for metamathematics in general lies in the fact that recursive definition enables every number in a recursively defined infinite sequence to be *constructed* according to a rule, so that a remark about the infinite sequence can be construed as a remark about the rule of construction and not as a remark about a given infinite totality. For this reason the use of only such mathematical concepts as are recursive is favoured by mathematical thinkers of both the finitist and intuitionist schools of metamathematics, and is accepted (although with extensions made by Gödel and others to the notion of recursiveness in this paper) by present-day constructivists who decline to talk about any mathematical entities that cannot be recursively constructed.

For the proof of Gödel's 'Unprovability' Theorem the importance of recursiveness lies in the fact (Proposition V, p. 55) that every statement of a recursive relationship holding between given numbers $x_1, x_2, \ldots x_n$ is expressible by a formula f of the formal system P which is 'provable' within P if the statement is true and 'disprovable' within P (i.e. the 'negation' of f, written as Neg f, is 'provable' within P) if the statement is false. Gödel only outlines a proof of this proposition, since it "offers no difficulties of principle and is somewhat involved" (p. 56); so I will expand what he says in his footnote (p. 56, n. 39). Since the relation R in question is recursive, then if $R(x_1, x_2, \ldots x_n)$ is true, $R(x_1, x_2, \ldots x_n)$ can be proved in a *deductive system for arithmetic* by constructing a finite sequence of propositions starting with the axioms and ending with $R(x_1, x_2, \ldots x_n)$; and if $R(x_1, x_2, \ldots x_n)$ is false, Not $R(x_1, x_2, \ldots x_n)$ can similarly be proved. The *calculus* or formal system P was designed by Gödel to represent this deductive system; so the finite sequence of propositions which constitutes a proof of $R(x_1, x_2, \ldots x_n)$ or of Not $R(x_1, x_2, \ldots x_n)$ will be expressed in P by a finite series of formulae ending in a formula f in the one case and in the formula Neg f in the other. To express in P the step-by-step definitional procedure by which the truth or falsity of the recursive relationship is established is to construct either a 'proof' of f or a 'proof' of Neg f: f or Neg f will only appear in the formal system accompanied by a 'proof-schema' of which it is the last formula. So if $R(x_1, x_2, \ldots x_n)$ is true, there is a 'proof' of f, and f is a 'provable' formula (definition 46, p. 55); and if $R(x_1, x_2, \ldots x_n)$ is false, Neg f is a 'provable' formula within the system P.

Define a *class-sign* as a series of signs which is a formula and which contains exactly one free variable (which may occur at several places in the formula) (p. 43). [In Gödel's system P there is no distinction between a class-sign and a property-sign, since 'axiom-schema' V (p. 44) may be regarded as expressing the axioms that two properties (of the same type) which always go together are identical— "axioms of extensionality".] A class-sign is *recursive* if it can be interpreted as expressing a recursive arithmetical class, in which case the formula resulting from the substitution for its variable of a number-sign will be 'provable' or 'disprovable' according as the number represented by the number-sign in the interpretation of the system is or is not a member of this recursive class. A *recursive relation-sign* is defined similarly (p. 43: see also p. 47, n. 28). Note that neither a class-sign nor a relation-sign is a basic sign, since the former contains one and the latter several free variables.

THE 'UNPROVABILITY' THEOREM FOR P. "We now come", as Gödel says (p. 56), "to the object of our exercises"—the proof of the 'Unprovability' Theorem. To prove this he establishes Proposition VI which is more general than is necessary for proving that there are undecidable formulae in the formal system P, since it is concerned not only with 'proofs' within P but also with 'deductions' within P from formulae not included among the 'axioms' of P, i.e. with 'proofs' within a formal system P′ obtained from P by adding these formulae as additional 'axioms'. Gödel requires this subtlety later on in his paper; but it complicates the proof of Proposition VI, which I shall discuss in the simplified form in which the class c of added formulae is the null class (i.e. no formulae are added to the axioms), so that a 'c-provable' formula within P (p. 59) is the same as a 'provable' formula within P, and the argument is concerned solely with 'proofs' within P. Gödel's B_c is thus taken as equivalent to his relation B, and Bew_c as equivalent to his property Bew.

Proposition VI simplified in this way may be stated as follows: *If the formal system* P *satisfies a certain condition of 'consistency', then there is at least one recursive class-sign* r *in* P *such that neither* v Gen r *nor* Neg (v Gen r) *is 'provable' within* P, *where* v Gen r *is the generalization of r with respect to its free variable* v.

The undecidability of v Gen r within P depends upon P's satisfying a certain 'consistency' condition. Since this condition is only relevant to the last stage of the proof, and itself raises important questions, consideration of 'consistency' will be deferred until the main part of the proof has been discussed.

This main part is given in pp. 58f. from (8.1) to (16). Gödel states his argument in terms of Gödel numbers and of relations between Gödel numbers; and when the expressions *relation-sign, free variable, class-sign, provable* are used they are italicized to show that they refer to arithmetical concepts applicable to Gödel numbers. Because of the correspondence between these concepts as applied to Gödel numbers and metamathematical concepts as applied to the strings which have these Gödel numbers, Gödel's whole argument applies equally well if his symbols are interpreted as strings and the terms relation-sign, free variable, etc. are taken in their usual sense. Since Gödel's argument, though couched in terms of numbers, is a metamathematical argument, it may be convenient for philosophical logicians if I give it wholly in metamathematical terms. This will have the additional advantage that interpretations of the formulae can be inserted parenthetically at appropriate places, on the assumption that the calculus (Gödel's formal system P) is to be interpreted as representing a deductive system which includes propositional and first-order predicate logic—though, strictly speaking, any actual interpretation is irrelevant to the argument.

However, a recasting of Gödel's argument in meta-mathematical terms makes one unimportant modification necessary. For the part of his argument which establishes the 'unprovability' of Neg (v Gen r) requires at one point considering a statement about all numbers, whether or not they are Gödel numbers; and this statement cannot be construed without change as a statement about all strings, since a number which is not a Gödel number does not correspond to any string. But it is easy to close the gap in the recasting by considering the numbers which are Gödel numbers as arranged in a sequence of increasing magnitude, and then using, instead of a Gödel number itself, the number which gives the place of this Gödel number in the sequence. To be precise, if n is the $(m+1)$-th Gödel number in increasing order, call m the G-*number* of the string of which n is the Gödel number, and use the G-number m wherever Gödel in his argument uses the Gödel number n. Then every natural number 0, 1, 2, etc. will be the G-number of some string; and there will be a recursive one-to-one correspondence between natural numbers and strings. So arithmetical statements about all numbers can be construed as metamathematical statements about all strings. Of course Gödel's sequence of definitions 6-46 defines arithmetical concepts which correspond to metamathematical concepts according to the Gödel-number method of arithmetization. But the purpose of his definitions is to establish that all the arithmetical concepts concerned (except Bew) are recursive and so are also all the corresponding metamathematical concepts. Consequently any proposition about them is expressible in P by a formula which is 'provable' or 'disprovable' according as the proposition is true or false.

Having proved this (by Proposition V, p. 55) Gödel makes
no further use in his argument for the 'Unprovability'
Theorem of his particular method of arithmetization. All
that is necessary is that there should be a unique number
assigned to every string. So no harm will result from con-
tinuing the argument using G-numbers instead of the corresponding Gödel numbers; and this use of G-numbers I shall
call the "modified arithmetization".

To facilitate comparison with Gödel's text, I shall use
Gödel's symbols, except that, as well as single small italic
letters denoting numbers, I shall in future use the same
letters in bold type to stand for the strings of which these
numbers are G-numbers. Thus **x** will be the string whose
G-number is x. Gödel writes $Z(x)$ for the Gödel number of
the number-sign for the number x in his formal system P
(see definition 17, p. 51). This number-sign is "0" preceded
by x "f" s; e.g. the number-sign for 3 is "$fff\,0$" (see p. 42).
I shall call these number-signs *numerals*; and shall write
Gx (or, if **x** is a complex expression, **G[x]**) for the numeral
for the G-number of **x** and call **Gx** the G-*numeral* of **x**.
Since every number is a G-number, every numeral is a
G-numeral; and there is a recursive one-to-one correspondence between the members of the class of numerals
"0", "$f\,0$", "$f\!f\,0$", etc. and the members of the class of
strings (which, of course, includes the class of numerals as
a sub-class).

A class-sign will be written in the form **a(v)** and a dyadic
relation-sign in the form **b (v, w)** with **v** or **v, w**, the free
variables (of first type) concerned, mentioned explicitly. [But
the G-numerals of **a(v)** and of **b (v, w)** will be abbreviated to
Ga and to **Gb**.] Since we are concerned with the formal
system P whose "individuals" are natural numbers (p. 42)
the substitution values for **v** and **w** will always be numerals,
and thus always G-numerals. The result of substituting **Gx**

for **v** and Gy for **w** in **b** (**v**, **w**) will be written as **b** (Gx, Gy). [Gödel uses a typographically less convenient notation for substitution. In comparing my version with his text it should be remembered that 17 is the Gödel number of **v** and 19 that of **w**.]

The simplified Proposition VI may now be restated as: *If the formal system* P *satisfies a certain condition of 'consistency', then there is at least one recursive class-sign* **r**(**v**) *in* P *such that neither* **v** Gen **r**(**v**) *nor* Neg [**v** Gen **r**(**v**)] *is 'provable' within* P.

We can now follow the principal steps in the argument of pp. 57-59 from (8.1) onwards.

Define Q' (**x**, **y**(**u**)) as Not [**x** B **y** (Gy)], i.e. **x** is not a 'proof' of the formula obtained by substituting for the variable in the class-sign **y**(**u**) the G-numeral Gy for the class-sign itself.

Let Q (x, y) be the relationship between the G-numbers of **x** and of **y** which is equivalent to Q' (**x**, **y** (**u**)) by the modified arithmetization. Q (x, y) is recursive; and so it follows from Proposition V that there is a recursive relation-sign **q** (**v**, **w**) which is such that

$$\begin{cases} Q \ (x, y) \to [\mathbf{q} \ (\text{Gx, Gy})] \text{ is 'provable'}; \\ \text{Not } Q \ (x, y) \to [\text{Neg } \mathbf{q} \ (\text{Gx, Gy})] \text{ is 'provable'}. \end{cases}$$

But Q' (**x**, **y** (**u**)) is equivalent to Q (x, y); and thus

$$\begin{cases} Q' \ (\mathbf{x}, \mathbf{y} \ (\mathbf{u})) \to [\mathbf{q} \ (\text{Gx, Gy})] \text{ is 'provable'}; \\ \text{Not } Q' \ (\mathbf{x}, \mathbf{y} \ (\mathbf{u})) \to [\text{Neg } \mathbf{q} \ (\text{Gx, Gy})] \text{ is 'provable'}. \end{cases}$$

The relation-sign **q** (**v**, **w**) may therefore be regarded as a formula expressing the relation which **x** has to **y**(**u**) when **x** is not a 'proof' of **y**(Gy).

Consider the 'generalization' of the relation-sign **q** (**v**, **w**) with respect to the free variable **v**, yielding the formula **v** Gen **q** (**v**, **w**). This has one free variable, namely **w**, and so is a class-sign. Call it **p(w)**. It may be regarded as denoting the class of which a class-sign **y(u)** is a member if and only if everything is not a 'proof' of **y(Gy)**, i.e. if and only if **y(Gy)** is 'unprovable'.

Next consider the substitution in the same relation-sign **q (v, w)** of **Gp** for the free variable **w**, yielding the formula **q (v, Gp)**. This also has one free variable, **v**, and so is also a class-sign. Call it **r(v)**. It may be regarded as denoting the class of strings which are not 'proofs' of **p(Gp)**. Since it may also be regarded, according to the modified arithmetization, as denoting the class of the G-numbers of these strings, which is a recursive arithmetical class, **r(v)** is a recursive class-sign.

Now consider the 'generalization' of this class-sign **r(v)** i.e. of **q (v, Gp)**, with respect to its free variable **v**, which yields the formula **v Gen r(v)**. This has no free variable, and may be regarded as expressing the proposition that everything is not a 'proof' of **p(Gp)**, i.e. that **p(Gp)** is 'unprovable'.

But, and here is the crux of the argument, **v Gen r(v)** *is the same as* **p(Gp)**. For we arrived at the former by first substituting **Gp** for **w** in **q (v, w)**, which yielded **r(v)**, and then 'generalizing' with respect to **v**, which yielded **v Gen r(v)**. But, since the substitution and the 'generalization' had reference to different free variables, the two operations yield the same final result if performed in the reverse order, i.e. by first 'generalizing' **q (v, w)** with respect to **v**, which yields **p(w)**, and then substituting **Gp** for **w** in **p(w)**, which yields **p(Gp)**. If either of the formulae **v Gen r(v)** or **p(Gp)** be expanded to get rid of the abbreviations **r** and **p**, we get one and the same formula

v Gen **q** (v, G [v Gen **q** (v, w)]).

This formula, and of course each of the abbreviations of

it, may be regarded as expressing the proposition that the formula itself is 'unprovable', i.e. the formula expresses its own 'unprovability'.

The formula of Gödel's which I have sometimes quoted, namely "17 Gen r", is the modified arithmetization of my metamathematical "v Gen $r(v)$", but with 17, the Gödel number of my variable v, used instead of the G-number of v. Since it is immaterial in which way the metamathematical formula is written, I shall in the next few pages use the shortest form, namely $p(Gp)$.

Now for the last stages of the proof. We go back to Q' $(x, y(u))$, defined as Not $[x \; B \; y(Gy)]$, i.e. as expressing the metamathematical proposition that x is not a 'proof' of $y(Gy)$. If we take the class-sign $y(u)$ to be $p(u)$, which is the same as $p(w)$, since u and w are variables, we get for the consequences of the truth or falsity of Q' $(x, p(u))$, i.e. of the truth or falsity of Not $[x \; B \; p(Gp)]$:

$$\begin{cases} \text{Not } [x \; B \; p(Gp)] \rightarrow [q \; (Gx, Gp)] \text{ is 'provable'}; \\ \quad x \; B \; p(Gp) \rightarrow [\text{Neg } q \; (Gx, Gp)] \text{ is 'provable'}. \end{cases}$$

q (Gx, Gp) is the same as $r(Gx)$ (which corresponds to Gödel's expression in square brackets on the right-hand side of (15), p. 59).

Suppose now that $p(Gp)$ were to be 'provable'. Then there would be a 'proof-schema' n such that $n \; B \; p(Gp)$, and hence such that Neg q (Gn, Gp) i.e. Neg $r(Gn)$ would be 'provable'. But if $p(Gp)$, i.e. v Gen $r(v)$, were 'provable', so also would be $r(Gn)$. So from the supposition that $p(Gp)$ is 'provable', there follows that both $r(Gn)$ and Neg $r(Gn)$ are 'provable'. Call a formal system (a calculus) *'consistent'* if it contains no pair of 'provable' formulae of the forms f, Neg f. Then, if $p(Gp)$ is 'provable', the formal system P is 'inconsistent'; so, if P is 'consistent', $p(Gp)$ is 'unprovable' within P.

Suppose that P is 'consistent' and that Neg p(Gp) were to be 'provable'. Since p(Gp) is 'unprovable', Not [n B p(Gp)] holds for every string n. Thus q (Gn, Gp), i.e. r(Gn), is 'provable' for every string n; and hence r(m) is 'provable' for every numeral m. But Neg p(Gp) is the same as Neg [v Gen r(v)]; and, if this were to be 'provable', the curious situation would arise of every substitution-instance r(m) of the class-sign r(v) being 'provable' while the 'generalization' of r(v) with respect to v was 'disprovable'. This situation is, however, compatible with the 'consistency' of P: in order to prohibit its occurrence a stronger form of consistency, called by Gödel 'ω-consistency', must be assumed to hold of P. [A formal system is 'ω-consistent' if it contains no class-sign a(u) which is such both that a(m) is 'provable' within the system for every numeral m and that Neg [u Gen a(u)] is 'provable' within the system (see p. 57).] Since the 'consistency' (sometimes called 'simple consistency') of P is a consequence of its 'ω-consistency' (see p. 23), the conjunction of 'ω-consistency' with the 'provability' of Neg p(Gp) yields a contradiction; so, if the formal system P is 'ω-consistent', Neg p(Gp) is 'unprovable' within P.

Combining these two results tells us that, if P is 'ω-consistent', neither p(Gp) nor Neg p(Gp) is 'provable' within P, i.e. p(Gp) is undecidable within P.

In order to compare this with my simplified restatement of Gödel's Proposition VI (p. 18) we must remember that p(Gp) is the same as v Gen r(v). r(v) is a recursive class-sign; so there is a recursive class-sign r(v) in P such that neither v Gen r(v) nor Neg [v Gen r(v)] is 'provable' within P, if P is 'ω-consistent'.

'CONSISTENCY'. If a formal system (a calculus) is 'inconsistent', it will contain both a 'provable' formula f and a 'provable' formula Neg **f**. If its 'axioms' and 'rules of inference' are such that the 'inconsistent' calculus can be interpreted, with Neg interpreted as meaning Not and a 'provable' formula interpreted as standing for a provable proposition, so that the deductive system which it represents includes a deductive sub-system of propositional logic (as is the case with Gödel's calculus P), then this deductive system will have as theorems both a proposition p (namely, the proposition represented by **f**) and its contradictory Not p (the proposition represented by Neg **f**), and hence the conjunction p & Not p—a self-contradiction. The deductive system will thus be inconsistent in the usual sense of the term, which, of course, is why Gödel uses the same word to apply to a calculus and I use the same word within single quotation marks.

Since $p \supset (p \lor q)$, which is equivalent to $(p$ & Not $p) \supset q$, is either an axiom or a theorem in propositional logic (the formula representing it in Gödel's P falls under 'axiom-schema' II.2: see p. 44), and since *modus ponens* (q is an immediate consequence of p and $p \supset q$) is a rule of inference in propositional logic (Gödel's P uses the corresponding 'rule': see p. 45), every proposition is a consequence of a self-contradiction. An inconsistent deductive system which includes a sub-system of propositional logic will therefore contain every proposition whatever as a theorem of the system. So if a calculus can be interpreted as representing a deductive system with the very small number of features required for it to include propositional logic, and if the calculus is 'inconsistent', every (well-formed) formula in the calculus will be 'provable' within the calculus. Such a calculus will be of no interest, since there will be no division of its formulae into those which are 'provable' and those which are not. This is the principal reason why meta-mathematicians attach such importance to a calculus being 'consistent', altogether apart from whether or not the calculus is in fact interpreted to represent a deductive system.

If the calculus P were 'inconsistent', all its formulae would be 'provable' and so the condition for 'ω-consistency' (p. 21) could not be satisfied. So if P is 'ω-consistent', it is also 'consistent'. The notion of 'ω-consistency' is intimately connected with finitist methods of proof. It will not be further considered here, since it is not a necessary condition in an 'Unprovability' Theorem for Gödel's formal system P. In 1936 Rosser, by an argument involving a recursive class-sign more complicated than Gödel's $r(v)$, established an 'Unprovability' Theorem for P (and for systems of similar character) which required as a condition only that P is 'consistent'.

A principal aim of Hilbert and his school had been to establish the 'consistency' of a calculus capable of being interpreted as expressing arithmetic, and thus to prove the consistency of a deductive system of arithmetic. To them the second great theorem contained in this paper was even more of a shock than the 'Unprovability' Theorem. For this second theorem proves the undecidability within P of a formula expressing the 'consistency' of P, thus showing that the 'consistency' of P, if P is 'consistent', cannot be established by a 'proof' within P, i.e. a 'proof' starting with only the 'axioms' of P and using only P's 'rules of inference'. [If P is 'inconsistent', of course both P's 'consistency' and P's 'inconsistency' can be 'proved' within P.]

THE 'UNPROVABILITY'-OF-'CONSISTENCY' THEOREM FOR P. Gödel proves this theorem (his Proposition XI: p. 70) in a general form, corresponding to that of his Proposition VI, which is concerned with 'deductions' as well as 'proofs' within P. As with Proposition VI I shall discuss Proposition XI in a simplified form in which it is concerned solely with 'proofs' within P. The simplified form is obtained by taking the class c to be the null class, and consequently B_c and Bew_c as equivalent to B and to Bew respectively.

Proposition XI simplified in this way may be stated as follows: *If the formal system P is 'consistent', its 'consistency' is 'unprovable' within P.*

In order to prove this theorem Gödel uses the result established towards the end of the proof of his 'Unprovability' Theorem, namely that, if P is 'consistent', the formula $p(Gp)$ is 'unprovable'. Since, as we have seen (p. 19), this formula may be regarded as expressing its own 'unprovability', the metamathematical proposition

P is 'consistent' $\rightarrow$ $p(Gp)$ is 'unprovable'

may be expressed within P by the formula, 'provable' within P,

$$w \text{ Imp } p(Gp),$$

where w (this symbol no longer being used as a variable) is a recursive formula expressing in P the 'consistency' of P, and u Imp v expresses in P the propositional schema Not a or b (see definition 32, p. 53). Then it follows from the definition of 'immediate consequence' (definition 43, p. 55) that, if w were to be 'provable', $p(Gp)$ would also be 'provable'. But if P is 'consistent', $p(Gp)$ is 'unprovable', and so also is w. Thus a formula w expressing the 'consistency' of P is 'unprovable' within P—on the assumption, of course, that P is 'consistent'.

In this paper Gödel only professed to "sketch in outline" the proof of his Proposition XI, and the sequel in which he intended to present it "in detail" he never published. Indeed the part of the detailed proof which establishes that w Imp $p(Gp)$ is 'provable' within P requires exhibiting a 'proof' within P of w Imp $p(Gp)$, and this is a lengthy and complicated business. However there is *prima facie* a gap in Gödel's "sketch" of the proof, namely how a recursive formula w which expresses the 'consistency' of P can be constructed in P; but this gap can easily be closed by an argument which I owe to Rosser. Let t be a particular formula which is 'provable' in P; e.g. one of the 'axioms' of P. If Neg t is also 'provable', P is 'inconsistent'. But, if P is 'inconsistent', every (well-formed) formula is 'provable' in P, and so Neg t is 'provable'. Thus the 'inconsistency' of P is logically equivalent to the 'provability' of Neg t, and the 'consistency' of P to the 'unprovability' of Neg t. So all that is required is a recursive formula in P expressing the 'unprovability' of Neg t, which is easy to provide. x B y (x is a 'proof' of the formula y) is a recursive relation-sign (definition 45, p. 55) with x and y as its free variables; hence Neg (x B Neg t) is a recursive class-sign, with x as its free variable, and x Gen [Neg (x B Neg t)] is a recursive formula which expresses in P the 'unprovability' of Neg t, which is equivalent to the 'consistency' of P. So the w of the proof in the last paragraph may be taken to be x Gen [Neg (x B Neg t)], in which case the proof (when given in detail) will fully satisfy the requirements of finitists and constructivists.

Gödel says at the end of his paper that his 'Unprovability'-of-'Consistency' Theorem represents "no contradiction of the formalistic standpoint of Hilbert. For this standpoint presupposes only the existence of a consistency proof effected by finite means, and there might conceivably be finite proofs which *cannot* be stated in P" (p. 71). This was a pious hope of Gödel's, made reasonable when he uttered it by the lack of precision in Hilbert's notion of a proof "effected by finite means". Clarification of this notion, to which this paper and later work of Gödel notably contributed, have explicated it in terms of the concept of recursiveness and of extensions of this concept; and it is now certain that, within any formal system using only such concepts and capable of expressing arithmetic, it is impossible to establish its own 'consistency' (if it is 'consistent'). In 1936 Gentzen was able to prove the 'consistency' of such a formal system, but only by using non-constructive methods of proof ("transfinite induction") which fall outside the constructive 'rules of inference' of the system. Gödel, in this paper which established his two great theorems by methods which are constructive in a precise sense, on the one hand showed the essential limitations imposed upon constructivist formal systems (which include all systems basing a calculus for arithmetic upon "mathematical induction"), and on the other hand displayed the power of constructivist methods for establishing metamathematical truths. To a philosophical logician it appears an even more remarkable feat to have been able to establish the internal undecidability of some arithmetical formulae than to provide (as Hilbert's school would have wished) a decision procedure for the whole of arithmetic.

THE SYNTACTICAL CHARACTER OF GÖDEL'S THEOREMS. In concluding this Introduction I wish to elaborate a point I have made several times in passing, namely that Gödel's two great theorems are metamathematical theorems about a calculus (his formal system P) and are not, in themselves, metamathematical theorems about a deductive system which is an interpretation of the calculus. However, theorems about deductive systems are immediate corollaries. The statement that there are arithmetical propositions which are neither provable nor disprovable within their arithmetical system (which at the beginning of this Introduction I called Gödel's Theorem *tout court*) is, for the deductive system for arithmetic represented by the calculus P, a corollary of the 'Unprovability' Theorem for P.

To appreciate this, consider the formula v Gen r(v) whose undecidability (subject, of course, to P being 'ω-consistent') was established by the proof of the 'Unprovability' Theorem. Interpret the class-sign r(v) as denoting the class of G-numbers of the strings which are not 'proofs' of p(Gp) —the second interpretation of r(v) mentioned on p. 19. This class of numbers, specified thus metamathematically, may also be specified arithmetically by modifying the arithmetization of *series of formulae, formula, proof* provided by definitions 1-45 (pp. 49ff.). So if 'generalizing' r(v) with respect to v is interpreted as stating that the class of numbers denoted by r(v) is the universal class, the formula v Gen r(v) will be interpreted as expressing the proposition that every number is a member of a certain arithmetically specified class—a straight arithmetical proposition (call it *g*). If the calculus P (assumed to be 'ω-consistent') is interpreted as representing a deductive system S for arithmetic (and it was devised so that it could represent that part of the *Principia Mathematica* deductive system required for arithmetic), with the 'axioms' and 'rules of inference' of P representing the axioms and rules of inference of S (and such an interpretation permits the interpretation of v Gen r(v) as expressing the arithmetical proposition *g*), then *g* will be neither provable nor disprovable by the methods of proof available in S, i.e. neither *g* nor Not *g* will be a theorem of S. [In Section 3 of this paper Gödel uses *arithmetical* in a more restricted sense than I have used it, and establishes that, even in this restricted sense, there will be arithmetical propositions undecidable in S.] The undecidability (the 'unprovability' and 'undisprovability') of v Gen r(v) within P is transferred to the deductive system S represented by P to yield the undecidability (the unprovability and undisprovability) of *g* within S. Similarly the 'unprovability' within P of the 'consistency' of P (assumed to be 'consistent') transfers to S to yield the unprovability within S of the consistency of S.

The undecidability of some arithmetical propositions within the deductive system S may be classed among the *syntactical* metamathematical characteristics of the system S (represented by the calculus P), for the reason that this undecidability derives from the undecidability of some formulae within the calculus P which represents S. Deductive systems, unlike calculi, have also *semantical* metamathematical characteristics; in particular their propositions have or lack the semantical property of *being true*—what Gödel in his introductory Section 1 calls being "correct as regards content" (*inhaltlich richtig*). Connecting the syntactical property of being provable with the semantical property of being true by taking every proposition provable within S (i.e. every axiom and theorem of S) to be true (see p. 41) gives an additional kick to the undecidability in S of *g*—by adding *that g is true*. For the correlation of arithmetical and metamathematical propositions effected by the modified arithmetization ensures that *g* will be true if and only if v Gen v(r) is 'unprovable' within P, i.e. if and only if *g* is unprovable within S. Hence if *g* were not true, *g* would be provable within S and so true—a contradiction. Consequently if the axioms and theorems of the deductive-system-for-arithmetic S are true (and this implies the consistency of S, for otherwise two propositions *p* and Not *p*, which cannot both be true, would both be theorems of S), then there is an arithmetical proposition, namely *g*, which is not provable within S (a syntactical characteristic) but which nevertheless is true (a semantical characteristic). This metamathematical argument which, combines semantical with syntactical considerations, establishes the truth of an arithmetical proposition which cannot be proved within S.

In his introductory Section 1 Gödel intermingles semantical with syntactical considerations in sketching a proof of the undecidability of g (which is the reason why I have seldom referred to this section in this Introduction). The distinction between what is syntactical and what semantical was not made explicitly until a year or two later (by Tarski, whose work included rigorously establishing unprovability theorems that were semantical); but it is implicit in Gödel's remark towards the end of Section 1 that "the exact statement of the proof [of the undecidability of g], which now follows, will have among others the task of substituting for the second of these assumptions [that every provable formula is also correct as regards content] a purely formal and much weaker one" (p. 41). Gödel's proof in Section 2 is a purely syntactical proof about a calculus (the formal system P) whose interpretation as a deductive system for arithmetic is, strictly speaking, irrelevant to his argument. It is true that Gödel explains arithmetization as a way of co-ordinating strings in his calculus with *natural numbers*, and he discusses recursive functions in terms of natural numbers (and I have followed him in speaking of numbers in both these contexts). But whenever he talks about numbers, and thus makes a remark which is *prima facie* about a deductive system rather than about a calculus, the remark is always a syntactical remark about the deductive system, and is therefore in essence a remark about the calculus which represents the system. For example, when Gödel says at the beginning of Section 2 that his formal system P has "numbers as individuals", and speaks of "variables of first type (for individuals, i.e. natural numbers including 0)" (p. 42), all that is relevant to his argument is that *numerals* are the only substitution values (not containing variables) permitted for his variables of first type. This is shown most clearly when Gödel specifies the substitution operation in connexion with his 'axiom-schema' III.1 (p. 44), which requires the substitution for a variable of first type of a sign of first type, which he has previously explained as being "a combination of signs of the form: $a, fa, ffa, fffa$, etc., where a is either "0" [in which case the sign is a *number-sign*] or a variable of first type" (p. 42; in Gödel's text 0 occurs instead of "0", but this would seem to be a misprint).

Gödel's 'arithmetization' of metamathematical concepts (as also my 'modified arithmetization') is in fact effected by correlating to each string x another string which is a numeral: there is no need to pass from a string x to this numeral by the indirect route of first moving to the Gödel number (or G-number) of x and then passing from this number to the numeral which expresses it in the calculus P. In the argument the equivalence, for example, between the metamathematical proposition about P stating that the string (the series of formulae) n is a 'proof' of the string (the formula) y and an arithmetical relationship between the G-numbers n and y of these strings may equally well be construed as an equivalence between the metamathematical proposition and the occurrence as a 'theorem' of P (i.e. as a 'provable' formula within P) of an appropriate 'recursive' 'arithmetical' formula f containing the strings (the numerals) Gn and Gy. [The requirement that f should be 'recursive' ensures that, if f is not a 'theorem' of P, Neg f is.] The peculiarity of the 'recursive' class-sign $r(v)$ of the 'Un-provability' Theorem is that, if there were to be a string n which was a 'proof' of v Gen $r(v)$, the 'recursive' 'arith-metical' formula Neg $r(Gn)$ would occur as a 'theorem' of P, whereas $r(Gn)$ would also appear as a 'theorem' of P as an 'immediate consequence' of a formula falling under 'axiom-schema' III.1 (p. 44) and of v Gen $r(v)$. In other words, if v Gen $r(v)$ were to be a 'theorem' (derived by a 'proof' n), r(Gn) would be a 'theorem' for a reason internal to the calculus, and Neg r(Gn) would be a 'theorem' for the reason that it was the 'recursive' formula whose occurrence as a 'theorem' was equivalent, according to the 'modified arithmetization', to n being a 'proof' of v Gen r(v).

In the last paragraph, where part of the proof of Gödel's 'Unprovability' Theorem has been restated in terms which either are used within the calculus P or are syntactical terms used to describe features of P, I have put single quotation marks round 'recursive', 'arithmetical', 'arithmetization', 'modified arithmetization' to indicate that these words are being used (like 'theorem', 'proof', 'provable', etc.) as calculus terms and not as deductive-system terms. The whole of Gödel's formal argument in this paper is syntactical: that he arithmetizes metamathematics instead of only 'arithmetizing' it is purely a matter of expository convenience. For his arithmetization is in terms of recursive arithmetical concepts, and by his Proposition V (see also p. 15) the question as to whether or not a recursive arithmetical relationship holds between *numbers* is equivalent to the syntactical question as to which of two 'recursive' formulae containing *numerals*, of the forms f, Neg f respectively, is a 'theorem' of the calculus P. [In my sketch (pp. 18-21) of Gödel's proof of the 'Unprovability' Theorem I have declined to follow him in using such terms as *formula*, *proof*, *class-sign* with an arithmetical interpretation; and I have, so far as was conveniently possible, employed G-numerals instead of G-numbers.]

Thus Gödel's two great theorems are theorems about his calculus P: they assert the 'unprovability' within P of certain well-formed formulae of P (on the assumption that P is 'ω-consistent' or 'consistent' respectively). Of course the interest to the learned world of the calculus P is that it can be regarded as representing a deductive system for arithmetic in which, therefore, there are undecidable arithmetical propositions. Though Gödel's formal proofs apply only to P, he indicates how similar proofs would apply to any calculus satisfying two very general conditions (p. 62), conditions so general that any calculus capable of expressing arithmetic can hardly fail to satisfy them. So this paper of Gödel's proclaimed the thesis, which has been clarified and confirmed by the work of subsequent metamathematicians, that no calculus can be devised in such a way that every arithmetical proposition is represented in it by a formula which is either 'provable' or 'disprovable' within the calculus, and therefore that any deductive system whatever for arithmetic will have the general syntactical characteristic of *not* containing either a proof or a disproof of every arithmetical proposition.

This syntactical fact about arithmetic is sometimes described by saying that arithmetic, in its very nature, is *incomplete*. Gödel's discovery of this incompleteness, presented in this paper, is one of the greatest and most surprising of the intellectual achievements of this century.

[I am much indebted to Dr T. J. Smiley for criticizing most helpfully the penultimate draft of this Introduction.

R. B. B.]

NOTE

The symbols Gödel uses for metamathematical concepts or their Gödel numbers are mainly abbreviations of German words. Although the concepts themselves are carefully defined in the text, the following alphabetical list of the more important of these symbols with their etymology may be helpful to the reader:

A	from	"Anzahl"	= number
Aeq	from	"Aequivalenz"	= equivalence
Ax	from	"Axiom"	= axiom
B	from	"Beweis"	= proof
Bew	from	"Beweisbar"	= provable
Bw	from	"Beweisfigur"	= proof-schema
Con	from	"Conjunktion"	= conjunction
Dis	from	"Disjunktion"	= disjunction
E	from	"Einklammern"	= include in brackets
Elf	from	"Elementarformel"	= elementary formula
Ex	from	"Existenz"	= existence
Fl	from	"unmittelbare Folge"	= immediate consequence
Flg	from	"Folgerungsmenge"	= set of consequences
Form	from	"Formel"	= formula
Fr	from	"frei"	= free
FR	from	"Reihe von Formeln"	= series of formulae
Geb	from	"gebunden"	= bound
Gen	from	"Generalisation"	= generalization
Gl	from	"Glied"	= term

Imp	from	"Implikation"	= implication
l	from	"Länge"	= length
Neg	from	"Negation"	= negation
Op	from	"Operation"	= operation
Pr	from	"Primzahl"	= prime number
Prim	from	"Primzahl"	= prime number
R	from	"Zahlenreihe"	= number series
Sb	from	"Substitution"	= substitution
St	from	"Stelle"	= place
Su	from	"Substitution"	= substitution
Th	from	"Typenerhöhung"	= type-lift
Typ	from	"Typ"	= type
Var	from	"Variable"	= variable
Wid	from	"Widerspruchsfreiheit"	= consistency
Z	from	"Zahlzeichen"	= number-sign

The only way in which the translation deviates from Gödel's symbolism is that, from p. 57 onwards, c is used to stand for the class which Gödel denotes by κ.

ON FORMALLY UNDECIDABLE
PROPOSITIONS
OF PRINCIPIA MATHEMATICA
AND RELATED SYSTEMS

by Kurt Gödel, Vienna

ON FORMALLY UNDECIDABLE PROPOSITIONS
OF PRINCIPIA MATHEMATICA AND RELATED
SYSTEMS I[1]

by Kurt Gödel, Vienna

1

The development of mathematics in the direction of greater exactness has—as is well known—led to large tracts of it becoming formalized, so that proofs can be carried out according to a few mechanical rules. The most comprehensive formal systems yet set up are, on the one hand, the system of Principia Mathematica (PM)[2] and, on the other, the axiom system for set theory of Zermelo-Fraenkel (later extended by J. v. Neumann).[3] These two systems are so extensive that all methods of proof used in mathematics today have been formalized in them, i.e. reduced to a few axioms and rules of inference. It may therefore be surmised that these axioms and rules of inference are also sufficient

---

[1] Cf. the summary of the results of this work, published in *Anzeiger der Akad. d. Wiss. in Wien* (math.-naturw. Kl.) 1930, No. 19.

[2] A. Whitehead and B. Russell, *Principia Mathematica*, 2nd edition, Cambridge 1925. In particular, we also reckon among the axioms of PM the axiom of infinity (in the form: there exist denumerably many individuals), and the axioms of reducibility and of choice (for all types).

[3] Cf. A. Fraenkel, 'Zehn Vorlesungen über die Grundlegung der Mengenlehre', *Wissensch. u. Hyp.*, Vol. XXXI; J. v. Neumann, 'Die Axiomatisierung der Mengenlehre', *Math. Zeitschr.* 27, 1928, *Journ. f. reine u. angew. Math.* 154 (1925), 160 (1929). We may note that in order to complete the formalization, the axioms and rules of inference of the logical calculus must be added to the axioms of set-theory given in the above-mentioned papers. The remarks that follow also apply to the formal systems presented in recent years by D. Hilbert and his colleagues (so far as these have yet been published). Cf. D. Hilbert, *Math. Ann.* 88, *Abh. aus d. math. Sem. der Univ. Hamburg* I (1922), VI (1928); P. Bernays, *Math. Ann.* 90; J. v. Neumann, *Math. Zeitschr.* 26 (1927); W. Ackermann, *Math. Ann.* 93.

to decide *all* mathematical questions which can in any way at all be expressed formally in the systems concerned. It is shown below that this is not the case, and that in both the systems mentioned there are in fact relatively simple problems in the theory of ordinary whole numbers[4] which cannot be decided from the axioms. This situation is not due in some way to the special nature of the systems set up, but holds for a very extensive class of formal systems, including, in particular, all those arising from the addition of a finite number of axioms to the two systems mentioned,[5] provided that thereby no false propositions of the kind described in footnote 4 become provable.

Before going into details, we shall first indicate the main lines of the proof, naturally without laying claim to exactness. The formulae of a formal system—we restrict ourselves here to the system PM—are, looked at from outside, finite series of basic signs (variables, logical constants and brackets or separation points), and it is easy to state precisely just *which* series of basic signs are meaningful formulae and which are not.[6] Proofs, from the formal standpoint, are likewise nothing but finite series of formulae (with certain specifiable characteristics). For metamathematical purposes it is naturally immaterial what objects are taken as basic signs, and we propose to use natural numbers[7] for them. Accordingly, then, a formula is a finite

[174]

---

[4] I.e., more precisely, there are undecidable propositions in which, besides the logical constants $^-$ (not), $\vee$ (or), (x) (for all) and $=$ (identical with), there are no other concepts beyond $+$ (addition) and $.$ (multiplication), both referred to natural numbers, and where the prefixes (x) can also refer only to natural numbers.

[5] In this connection, only such axioms in PM are counted as distinct as do not arise from each other purely by change of type.

[6] Here and in what follows, we shall always understand the term "formula of PM" to mean a formula written without abbreviations (i.e. without use of definitions). Definitions serve only to abridge the written text and are therefore in principle superfluous.

[7] I.e. we map the basic signs in one-to-one fashion on the natural numbers (as is actually done on p. 45).

series of natural numbers,[8] and a particular proof-schema is a finite series of finite series of natural numbers. Meta-mathematical concepts and propositions thereby become concepts and propositions concerning natural numbers, or series of them,[9] and therefore at least partially expressible in the symbols of the system PM itself. In particular, it can be shown that the concepts, "formula", "proof-schema", "provable formula" are definable in the system PM, i.e. one can give[10] a formula $F(v)$ of PM—for example—with one free variable v (of the type of a series of numbers), such that $F(v)$—interpreted as to content—states: v is a provable formula. We now obtain an undecidable proposition of the system PM, i.e. a proposition A, for which neither A nor *not-A* are provable, in the following manner:

A formula of PM with just one free variable, and that of   [175]
the type of the natural numbers (class of classes), we shall designate a **class-sign**. We think of the class-signs as being somehow arranged in a series,[11] and denote the n-th one by $R(n)$; and we note that the concept "class-sign" as well as the ordering relation R are definable in the system PM. Let α be any class-sign; by $[\alpha; n]$ we designate that formula which is derived on replacing the free variable in the class-sign α by the sign for the natural number n. The three-term relation $x = [y; z]$ also proves to be definable in PM. We now define a class K of natural numbers, as follows:

[8] I.e. a covering of a section of the number series by natural numbers. (Numbers cannot in fact be put into a spatial order.)

[9] In other words, the above-described procedure provides an iso-morphic image of the system PM in the domain of arithmetic, and all metamathematical arguments can equally well be conducted in this isomorphic image. This occurs in the following outline proof, i.e. "formula", "proposition", "variable", etc. **are always to be understood as the corresponding objects of the isomorphic image.**

[10] It would be very simple (though rather laborious) actually to write out this formula.

[11] Perhaps according to the increasing sums of their terms and, for equal sums, in alphabetical order.

$$n \ \varepsilon \ K \ \equiv \ \overline{Bew} \ [R(n); n]^{11a} \tag{1}$$

(where *Bew x* means: *x* is a provable formula). Since the concepts which appear in the definiens are all definable in PM, so too is the concept *K* which is constituted from them, i.e. there is a class-sign S,[12] such that the formula [*S*; *n*]— interpreted as to its content—states that the natural number *n* belongs to *K*. *S*, being a class-sign, is identical with some determinate *R(q)*, i.e.

$$S = R(q)$$

holds for some determinate natural number *q*. We now show that the proposition $[R(q); q]$[13] is undecidable in PM. For supposing the proposition [*R(q)*; *q*] were provable, it would also be correct; but that means, as has been said, that *q* would belong to *K*, i.e. according to (1), $\overline{Bew} \ [R(q); q]$ would hold good, in contradiction to our initial assumption. If, on the contrary, the negation of [*R(q)*; *q*] were provable, then $\overline{n \ \varepsilon \ K}$, i.e. *Bew* [*R(q)*; *q*] would hold good. [*R(q)*; *q*] would thus be provable at the same time as its negation, which again is impossible.

The analogy between this result and Richard's antinomy leaps to the eye; there is also a close relationship with the "liar" antinomy,[14] since the undecidable proposition [*R(q)*; *q*] states precisely that *q* belongs to *K*, i.e. according to (1), that [*R(q)*; *q*] is not provable. We are therefore confronted with a proposition which asserts its own unprov-

[11a] The bar-sign indicates negation.

[12] Again there is not the slightest difficulty in actually writing out the formula *S*.

[13] Note that "[*R(q)*; *q*]" (or—what comes to the same thing— "[*S*; *q*]") is merely a **metamathematical description** of the undecidable proposition. But as soon as one has ascertained the formula *S*, one can naturally also determine the number *q*, and thereby effectively write out the undecidable proposition itself.

[14] Every epistemological antinomy can likewise be used for a similar undecidability proof.

ability.[15] The method of proof just exhibited can clearly be applied to every formal system having the following [176] features: firstly, interpreted as to content, it disposes of sufficient means of expression to define the concepts occurring in the above argument (in particular the concept "provable formula"); secondly, every provable formula in it is also correct as regards content. The exact statement of the above proof, which now follows, will have among others the task of substituting for the second of these assumptions a purely formal and much weaker one.

From the remark that $[R(q); q]$ asserts its own unprovability, it follows at once that $[R(q); q]$ is correct, since $[R(q); q]$ is certainly unprovable (because undecidable). So the proposition which is undecidable *in the system* PM yet turns out to be decided by metamathematical considerations. The close analysis of this remarkable circumstance leads to surprising results concerning proofs of consistency of formal systems, which are dealt with in more detail in Section 4 (Proposition XI).

2

We proceed now to the rigorous development of the proof sketched above, and begin by giving an exact description of the formal system P, for which we seek to demonstrate the existence of undecidable propositions. P is essentially the system obtained by superimposing on the Peano axioms the logic of PM[16] (numbers as individuals, relation of successor as undefined basic concept).

[15] In spite of appearances, there is nothing circular about such a proposition, since it begins by asserting the unprovability of a wholly determinate formula (namely the q-th in the alphabetical arrangement with a definite substitution), and only subsequently (and in some way by accident) does it emerge that this formula is precisely that by which the proposition was itself expressed.

[16] The addition of the Peano axioms, like all the other changes made in the system PM, serves only to simplify the proof and can in principle be dispensed with.

42   KURT GÖDEL

The basic signs of the system P are the following:

I. Constants: "$\sim$" (not), "$\vee$" (or), "Π" (for all), "0" (nought), "f" (the successor of), "(", ")" (brackets).

II. Variables of first type (for individuals, i.e. natural numbers including 0): "x_1", "y_1", "z_1", ...

Variables of second type (for classes of individuals): "x_2", "y_2", "z_2", ...

Variables of third type (for classes of classes of individuals): "x_3", "y_3", "z_3", ...

and so on for every natural number as type.[17]

Note: Variables for two-termed and many-termed functions (relations) are superfluous as basic signs, since relations can be defined as classes of ordered pairs and ordered pairs again as classes of classes, e.g. the ordered pair a, b by $((a), (a, b))$, where (x, y) means the class whose only elements are x and y, and (x) the class whose only element is x.[18]

[177]   By a **sign of first type** we understand a combination of signs of the form:

$$a, fa, ffa, fffa \ldots \text{etc.}$$

where a is either 0 or a variable of first type. In the former case we call such a sign a **number-sign**. For $n > 1$ we understand by a **sign of n-th type** the same as **variable of n-th type**. Combinations of signs of the form $a(b)$, where b is a sign of n-th and a a sign of $(n+1)$-th type, we call **elementary**

[17] It is presupposed that for every variable type denumerably many signs are available.

[18] Unhomogeneous relations could also be defined in this manner, e.g. a relation between individuals and classes as a class of elements of the form: $((x_2), ((x_1), x_2))$. As a simple consideration shows, all the provable propositions about relations in PM are also provable in this fashion.

formulae. The class of **formulae** we define as the smallest class[19] containing all elementary formulae and, also, along with any a and b the following: $\sim(a)$, $(a) \vee (b)$, $x \Pi (a)$ (where x is any given variable).[18a] We term $(a) \vee (b)$ the **disjunction** of a and b, $\sim(a)$ the **negation** and $x \Pi (a)$ a **generalization** of a. A formula in which there is no free variable is called a **propositional formula** (**free variable** being defined in the usual way). A formula with just n free individual variables (and otherwise no free variables) we call an **n-place relation-sign** and for $n = 1$ also a **class-sign**.

By Subst $a\begin{pmatrix} v \\ b \end{pmatrix}$ (where a stands for a formula, v a variable and b a sign of the same type as v) we understand the formula derived from a, when we replace v in it, wherever it is free, by b.[20] We say that a formula a is a **type-lift** of another one b, if a derives from b, when we increase by the same amount the type of all variables appearing in b.

The following formulae (I-V) are called **axioms** (they are set out with the help of the customarily defined abbreviations: $.$, $\supset$, $\equiv$, (Ex), $=$,[21] and subject to the usual conventions about omission of brackets)[22]:

[18a] Thus $x \Pi (a)$ is also a formula if x does not occur, or does not occur free, in a. In that case $x \Pi (a)$ naturally means the same as a.

[19] With regard to this definition (and others like it occurring later), cf. J. Lukasiewicz and A. Tarski, 'Untersuchungen über den Aussagenkalkül', *Comptes Rendus des séances de la Société des Sciences et des Lettres de Varsovie* XXIII, 1930, Cl. III.

[20] Where v does not occur in a as a free variable, we must put Subst $a\binom{v}{b} = a$. Note that "Subst" is a sign belonging to metamathematics.

[21] As in PM I, $*13$, $x_1 = y_1$ is to be thought of as defined by $x_2 \Pi (x_2(x_1) \supset x_2(y_1))$ (and similarly for higher types.)

[22] To obtain the axioms from the schemata presented (and in the cases of II, III and IV, after carrying out the permitted substitutions), one must therefore still
  1. eliminate the abbreviations,
  2. add the suppressed brackets.
Note that the resultant expressions must be "formulae" in the above sense. (Cf. also the exact definitions of the metamathematical concepts on pp. 49ff.)

I.　　　1. $\sim (fx_1 = 0)$

　　　　2. $fx_1 = fy_1 \supset x_1 = y_1$

　　　　3. $x_2(0) \, . \, x_1 \Pi (x_2(x_1) \supset x_2(fx_1)) \supset x_1 \Pi (x_2(x_1))$

[178] II. Every formula derived from the following schemata by substitution of any formulae for p, q and r.

　　　1. $p \vee p \supset p$　　　3. $p \vee q \supset q \vee p$

　　　2. $p \supset p \vee q$　　　4. $(p \supset q) \supset (r \vee p \supset r \vee q)$

III. Every formula derived from the two schemata

$$1. \; v \Pi (a) \supset \text{Subst } a \begin{pmatrix} v \\ c \end{pmatrix}$$

$$2. \; v \Pi (b \vee a) \supset b \vee v \Pi (a)$$

by making the following substitutions for a, v, b, c (and carrying out in 1 the operation denoted by "Subst"): for a any given formula, for v any variable, for b any formula in which v does not appear free, for c a sign of the same type as v, provided that c contains no variable which is bound in a at a place where v is free.[23]

IV. Every formula derived from the schema

$$1. \; (Eu) \, (v \Pi (u(v) \equiv a))$$

on substituting for v or u any variables of types n or $n+1$ respectively, and for a a formula which does not contain u free. This axiom represents the axiom of reducibility (the axiom of comprehension of set theory).

V. Every formula derived from the following by type-lift (and this formula itself):

$$1. \; x_1 \Pi (x_2(x_1) \equiv y_2(x_1)) \supset x_2 = y_2.$$

[23] c is therefore either a variable or 0 or a sign of the form $f \ldots fu$ where u is either 0 or a variable of type 1. With regard to the concept "free (bound) at a place in a" cf. section I A5 of the work cited in footnote 24.

This axiom states that a class is completely determined by its elements.

A formula c is called an **immediate consequence** of a and b, if a is the formula $(\sim(b)) \vee (c)$, and an **immediate consequence** of a, if c is the formula $v \, \Pi \, (a)$, where v denotes any given variable. The class of **provable formulae** is defined as the smallest class of formulae which contains the axioms and is closed with respect to the relation "immediate consequence of".[24]

The basic signs of the system P are now ordered in one-to-one correspondence with natural numbers, as follows:

"0" . . . 1	"∨" . . . 7	"(" . . . 11		[179]
"f" . . . 3	"Π" . . . 9	")" . . . 13		
"∼" . . . 5				

Furthermore, variables of type n are given numbers of the form p^n (where p is a prime number >13). Hence, to every finite series of basic signs (and so also to every formula) there corresponds, one-to-one, a finite series of natural numbers. These finite series of natural numbers we now map (again in one-to-one correspondence) on to natural numbers, by letting the number $2^{n_1} \cdot 3^{n_2} \ldots p_k^{n_k}$ correspond to the series $n_1, n_2, \ldots n_k$, where p_k denotes the k-th prime number in order of magnitude. A natural number is thereby assigned in one-to-one correspondence, not only to every basic sign, but also to every finite series of such signs. We denote by $\Phi(a)$ the number corresponding to the basic sign or series of basic signs a. Suppose now one is given a class or relation $R(a_1, a_2, \ldots a_n)$ of basic signs or series of such. We assign to it that class (or relation)

---

[24] The rule of substitution becomes superfluous, since we have already dealt with all possible substitutions in the axioms themselves (as is also done in J. v. Neumann, 'Zur Hilbertschen Beweistheorie', *Math. Zeitschr.* 26, 1927).

$R'(x_1, x_2, \ldots x_n)$ of natural numbers, which holds for $x_1, x_2, \ldots x_n$ when and only when there exist $a_1, a_2, \ldots a_n$ such that $x_i = \Phi(a_i)$ $(i = 1, 2, \ldots n)$ and $R(a_1, a_2, \ldots a_n)$ holds. We represent by the same words in italics those classes and relations of natural numbers which have been assigned in this fashion to such previously defined metamathematical concepts as "variable", "formula", "propositional formula", "axiom", "provable formula", etc. The proposition that there are undecidable problems in the system P would therefore read, for example, as follows: There exist *propositional formulae a* such that neither *a* nor the *negation* of *a* are *provable formulae*.

We now introduce a parenthetic consideration having no immediate connection with the formal system P, and first put forward the following definition: A number-theoretic function[25] $\phi(x_1, x_2, \ldots x_n)$ is said to be **recursively defined** by the number-theoretic functions $\psi(x_1, x_2, \ldots x_{n-1})$ and $\mu(x_1, x_2, \ldots x_{n+1})$, if for all $x_2, \ldots x_n, k$[26] the following hold:

$$\phi(0, x_2, \ldots x_n) = \psi(x_2 \ldots x_n)$$
$$\phi(k+1, x_2, \ldots x_n) = \mu(k, \phi(k, x_2, \ldots x_n), x_2, \ldots x_n). \quad (2)$$

A number-theoretic function ϕ is called **recursive**, if there exists a finite series of number-theoretic functions $\phi_1, \phi_2, \ldots \phi_n$, which ends in ϕ and has the property that every function ϕ_k of the series is either recursively defined by two of the earlier ones, or is derived from any of the earlier ones by substitution,[27] or, finally, is a constant or

[180]

[25] I.e. its field of definition is the class of non-negative whole numbers (or *n*-tuples of such), respectively, and its values are non-negative whole numbers.

[26] In what follows, small italic letters (with or without indices) are always variables for non-negative whole numbers (failing an express statement to the contrary).

[27] More precisely, by substitution of certain of the foregoing functions in the empty places of the preceding, e.g. $\phi_k(x_1, x_2) =$

the successor function $x + 1$. The length of the shortest series of ϕ_i, which belongs to a recursive function ϕ, is termed its **degree**. A relation $R(x_1 \ldots x_n)$ among natural numbers is called recursive,[28] if there exists a recursive function $\phi(x_1 \ldots x_n)$ such that for all $x_1, x_2, \ldots x_n$

$$R(x_1 \ldots x_n) \sim [\phi(x_1 \ldots x_n) = 0]^{29}.$$

The following propositions hold:

I. *Every function (or relation) derived from recursive functions (or relations) by the substitution of recursive functions in place of variables is recursive; so also is every function derived from recursive functions by recursive definition according to schema* (2).

II. *If R and S are recursive relations, then so also are $\bar{R}$, $R \vee S$ (and therefore also $R \And S$).*

III. *If the functions $\phi(\mathfrak{x})$ and $\psi(\mathfrak{y})$ are recursive, so also is the relation: $\phi(\mathfrak{x}) = \psi(\mathfrak{y})$.*[30]

IV. *If the function $\phi(\mathfrak{x})$ and the relation $R(x, \mathfrak{y})$ are recursive, so also then are the relations S, T*

$$S(\mathfrak{x}, \mathfrak{y}) \sim (Ex) [x \leqq \phi(\mathfrak{x}) \And R(x, \mathfrak{y})]$$
$$T(\mathfrak{x}, \mathfrak{y}) \sim (x) [x \leqq \phi(\mathfrak{x}) \rightarrow R(x, \mathfrak{y})]$$

$\phi_p [\phi_q(x_1, x_2), \phi_r(x_2)]$ $(p, q, r < k)$. Not all the variables on the left-hand side must also occur on the right (and similarly in the recursion schema (2)).

[28] We include classes among relations (one-place relations). Recursive relations R naturally have the property that for every specific n-tuple of numbers it can be decided whether $R(x_1 \ldots x_n)$ holds or not.

[29] For all considerations as to content (more especially also of a metamathematical kind) the Hilbertian symbolism is used, cf. Hilbert-Ackermann, *Grundzüge der theoretischen Logik*, Berlin 1928.

[30] We use gothic letters $\mathfrak{x}$, $\mathfrak{y}$, as abbreviations for given n-tuple sets of variables, e.g. $x_1, x_2 \ldots x_n$.

and likewise the function ψ

$$\psi(x, \mathfrak{y}) = \varepsilon x [x \leqq \phi(x) \And R(x, \mathfrak{y})],$$

where $\varepsilon x F(x)$ means: the smallest number x for which $F(x)$ holds and 0 if there is no such number.

Proposition I follows immediately from the definition of "recursive". Propositions II and III are based on the readily ascertainable fact that the number-theoretic functions corresponding to the logical concepts $\overline{}$, $\vee$, $=$

$$\alpha(x), \ \beta(x, y), \ \gamma(x, y)$$

namely

$$\alpha(0) = 1; \ \alpha(x) = 0 \quad \text{for} \quad x \neq 0$$
$$\beta(0, x) = \beta(x, 0) = 0; \ \beta(x, y) = 1, \text{ if } x, y \text{ both} \neq 0$$
$$\gamma(x, y) = 0, \text{ if } x = y; \quad \gamma(x, y) = 1, \text{ if } x \neq y$$

[181]

are recursive. The proof of Proposition IV is briefly as follows: According to the assumption there exists a recursive $\rho(x, \mathfrak{y})$ such that

$$R(x, \mathfrak{y}) \sim [\rho(x, \mathfrak{y}) = 0].$$

We now define, according to the recursion schema (2), a function $\chi(x, \mathfrak{y})$ in the following manner:

$$\chi(0, \mathfrak{y}) = 0$$
$$\chi(n+1, \mathfrak{y}) = (n+1) \cdot a + \chi(n, \mathfrak{y}) \cdot \alpha(a)^{31}$$

where

$$a = \alpha[\alpha(\rho(0, \mathfrak{y}))] \cdot \alpha[\rho(n+1, \mathfrak{y})] \cdot \alpha[\chi(n, \mathfrak{y})].$$

$\chi(n+1, \mathfrak{y})$ is therefore either $= n+1$ (if $a = 1$) or $= \chi(n, \mathfrak{y})$ (if $a = 0$).[32] The first case clearly arises if and only if all the constituent factors of a are 1, i.e. if

[31] We take it to be recognized that the functions $x + y$ (addition) and $x \cdot y$ (multiplication) are recursive.

[32] a cannot take values other than 0 and 1, as is evident from the definition of α.

$$\bar{R}(0, \mathfrak{y}) \,\&\, R\,(n+1, \mathfrak{y}) \,\&\, [\chi\,(n, \mathfrak{y}) = 0].$$

From this it follows that the function $\chi\,(n, \mathfrak{y})$ (considered as a function of n) remains 0 up to the smallest value of n for which $R\,(n, \mathfrak{y})$ holds, and from then on is equal to this value (if $R\,(0, \mathfrak{y})$ is already the case, the corresponding $\chi\,(x, \mathfrak{y})$ is constant and $=0$). Therefore:

$$\psi\,(\mathfrak{x}, \mathfrak{y}) = \chi\,(\phi\,(\mathfrak{x}), \mathfrak{y})$$
$$S\,(\mathfrak{x}, \mathfrak{y}) \sim R\,[\psi\,(\mathfrak{x}, \mathfrak{y}), \mathfrak{y}].$$

The relation T can be reduced by negation to a case analogous to S, so that Proposition IV is proved.

The functions $x+y$, $x \cdot y$, x^y, and also the relations $x<y$, $x=y$ are readily found to be recursive; starting from these concepts, we now define a series of functions (and relations) 1-45, of which each is defined from the earlier ones by means of the operations named in Propositions I to IV. This procedure, generally speaking, puts together many of the definition steps permitted by Propositions I to IV. Each of the functions (relations) 1-45, containing, for example, the concepts "*formula*", "*axiom*", and "*immediate consequence*", is therefore recursive.

1. $x/y \equiv (Ez)\,[z \leqq x \,\&\, x = y.z]$[33]   [182]
   x is divisible by y.[34]

2. $\mathrm{Prim}\,(x) \equiv \overline{(Ez)}\,[z \leqq x \,\&\, z \neq 1 \,\&\, z \neq x \,\&\, x/z]$
       $\&\, x > 1$
   x is a prime number.

[33] The sign $\equiv$ is used to mean "equivalence by definition", and therefore does duty in definitions either for $=$ or for $\sim$ (otherwise the symbolism is Hilbertian).

[34] Wherever in the following definitions one of the signs (x), $(E\,x)$, $\varepsilon\,x$ occurs, it is followed by a limitation on the value of x. This limitation merely serves to ensure the recursive nature of the concept defined. (Cf. Proposition IV.) On the other hand, the range of the defined concept would almost always remain unaffected by its omission.

3. $0 \, Pr \, x \equiv 0$

   $(n + 1) \, Pr \, x \equiv \varepsilon \, y \, [y \leqq x \, \& \, \text{Prim}(y) \, \& \, x/y$
   $\& \, y > n \, Pr \, x]$

   $n \, Pr \, x$ is the n-th (in order of magnitude) prime number contained in x.[34a]

4. $0! \equiv 1$

   $(n + 1)! \equiv (n + 1) \, . \, n!$

5. $Pr(0) \equiv 0$

   $Pr(n + 1) \equiv \varepsilon \, y \, [y \leqq \{Pr(n)\}! + 1 \, \& \, \text{Prim}(y)$
   $\& \, y > Pr(n)]$

   $Pr(n)$ is the n-th prime number (in order of magnitude).

6. $n \, Gl \, x \equiv \varepsilon \, y \, [y \leqq x \, \& \, x/(n \, Pr \, x)^y \, \& \, \overline{x/(n \, Pr \, x)^{y+1}}]$

   $n \, Gl \, x$ is the n-th term of the series of numbers assigned to the number x (for $n > 0$ and n not greater than the length of this series).

7. $l(x) \equiv \varepsilon \, y \, [y \leqq x \, \& \, y \, Pr \, x > 0 \, \& \, (y + 1) \, Pr \, x = 0]$

   $l(x)$ is the length of the series of numbers assigned to x.

8. $x * y \equiv \varepsilon \, z \, \big[z \leqq [Pr\{l(x) + l(y)\}]^{x+y}$
   $\& \, (n) \, [n \leqq l(x) \to n \, Gl \, z = n \, Gl \, x]$
   $\& \, (n) \, [0 < n \leqq l(y) \to \{n + l(x)\} \, Gl \, z = n \, Gl \, y] \big]$

   $x * y$ corresponds to the operation of "joining together" two finite series of numbers.

9. $R(x) \equiv 2^x$

   $R(x)$ corresponds to the number-series consisting only of the number x (for $x > 0$).

10. $E(x) \equiv R(11) * x * R(13)$

    $E(x)$ corresponds to the operation of "bracketing" [11 and 13 are assigned to the basic signs "(" and ")"].

[34a] For $0 < n \leqq z$, where z is the number of distinct prime numbers dividing into x. Note that for $n = z + 1$, $n \, Pr \, x = 0$.

11. $n \operatorname{Var} x \equiv (Ez) [13 < z \leqq x \ \& \ \operatorname{Prim}(z) \ \& \ x = z^n]$
    $\& \ n \neq 0$.
    x is a *variable of n-th type.*

12. $\operatorname{Var}(x) \equiv (En) [n \leqq x \ \& \ n \operatorname{Var} x]$
    x is a *variable.*

13. $\operatorname{Neg}(x) \equiv R(5) * E(x)$
    $\operatorname{Neg}(x)$ is the *negation* of x.

14. $x \operatorname{Dis} y \equiv E(x) * R(7) * E(y)$ [183]
    $x \operatorname{Dis} y$ is the *disjunction* of x and y.

15. $x \operatorname{Gen} y \equiv R(x) * R(9) * E(y)$
    $x \operatorname{Gen} y$ is the *generalization* of y by means of the
    variable x (assuming x is a *variable*).

16. $0 N x \equiv x$
    $(n + 1) N x \equiv R(3) * n N x$
    $n N x$ corresponds to the operation: "n-fold prefixing
    of the sign 'f' before x."

17. $Z(n) \equiv n N [R(1)]$
    $Z(n)$ is the *number-sign* for the number n.

18. $\operatorname{Typ}_1{}'(x) \equiv (Em, n) \{m, n \leqq x \ \& \ [m = 1 \lor 1 \operatorname{Var} m]$
    $\& \ x = n N [R(m)]\}^{34b}$
    x is a *sign of first type.*

19. $\operatorname{Typ}_n(x) \equiv [n = 1 \ \& \ \operatorname{Typ}_1{}'(x)] \lor [n > 1 \ \& \ (Ev) \{v \leqq x$
    $\& \ n \operatorname{Var} v \ \& \ x = R(v)\}]$
    x is a *sign of n-th type.*

20. $Elf(x) \equiv (Ey, z, n) [y, z, n \leqq x \ \& \ \operatorname{Typ}_n(y)$
    $\& \ \operatorname{Typ}_{n+1}(z) \ \& \ x = z * E(y)]$
    x is an *elementary formula.*

[34b] $m, n \leqq x$ stands for: $m \leqq x \ \& \ n \leqq x$ (and similarly for more than two variables).

21. $Op(x, y, z) \equiv x = \mathrm{Neg}(y) \lor x = y \mathrm{Dis} z \lor (Ev) [v \leqq x$
 $\& \mathrm{Var}(v) \& x = v \mathrm{Gen} y]$

22. $FR(x) \equiv (n) \{0 < n \leqq l(x) \to Elf(n Gl x) \lor (Ep, q)$
 $[0 < p, q < n \& Op (n Gl x, p Gl x, q Gl x)]\}$
 $\& l(x) > 0$

 x is a series of *formulae* of which each is either an
 elementary formula or arises from those preceding by
 the operations of *negation, disjunction* and *generalization.*

23. $\mathrm{Form}(x) \equiv (En) \{n \leqq (Pr [l(x)^2])^{x \cdot [l(x)]^2}$
 $\& FR(n) \& x = [l(n)] Gl n\}^{35}$

 x is a *formula* (i.e. last term of a *series of formulae n*).

24. $v \mathrm{Geb} n, x \equiv \mathrm{Var}(v) \& \mathrm{Form}(x) \& (Ea, b, c) [a, b, c \leqq x$
 $\& x = a \ast (v \mathrm{Gen} b) \ast c \& \mathrm{Form}(b)$
 $\& l(a) + 1 \leqq n \leqq l(a) + l(v \mathrm{Gen} b)]$

 The *variable v* is *bound* at the *n*-th place in *x*.

[184]

25. $v Fr n, x \equiv \mathrm{Var}(v) \& \mathrm{Form}(x) \& v = n Gl x$
 $\& n \leqq l(x) \& \overline{v \mathrm{Geb} n, x}$

 The *variable v* is *free* at the *n*-th place in *x*.

26. $v Fr x \equiv (En) [n \leqq l(x) \& v Fr n, x]$

 v occurs in *x* as a *free variable.*

27. $Su x \binom{n}{y} \equiv \varepsilon z \{z \leqq [Pr(l(x) + l(y))]^{x+y}$
 $\& [(Eu, v) u, v \leqq x \& x = u \ast R(n Gl x) \ast v$
 $\& z = u \ast y \ast v \& n = l(u) + 1]\}$

 $Su x \binom{n}{y}$ derives from *x* on substituting *y* in place of
 the *n*-th term of *x* (it being assumed that $0 < n \leqq l(x)$).

---

[35] The limitation $n \leqq (Pr [l(x)^2])^{x \cdot [l(x)]^2}$ means roughly this: The
length of the shortest series of formulae belonging to *x* can at most be
equal to the number of constituent formulae of *x*. There are however
at most $l(x)$ constituent formulae of length 1, at most $l(x) - 1$ of
length 2, etc. and in all, therefore, at most $\frac{1}{2} [l(x) \{l(x) + 1\}] \leqq [l(x)]^2$.
The prime numbers in *n* can therefore all be assumed smaller than
$Pr \{[l(x)]^2\}$, their number $\leqq [l(x)]^2$ and their exponents (which are constituent formulae of *x*) $\leqq x$.

28. $O\,St\,v,x \equiv \varepsilon\,n\,\{n \leqq l(x)\ \&\ v\,Fr\,n,x\ \&\ \overline{(Ep)}\,[n < p \leqq l(x)$
$\&\ v\,Fr\,p,x]\}$

$(k+1)\,St\,v,x \equiv \varepsilon\,n\,\{n < k\,St\,v,x$

$\&\ v\,Fr\,n,x\ \&\ (Ep)\,[n < p < k\,St\,v,x\ \&\ v\,Fr\,p,x]\}$

$k\,St\,v,x$ is the $(k+1)$th place in x (numbering from the end of the *formula* x) at which v is free in x (and 0, if there is no such place).

29. $A(v,x) \equiv \varepsilon\,n\,\{n \leqq l(x)\ \&\ n\ St\,v,x = 0\}$

$A(v,x)$ is the number of places at which v is *free* in x.

30. $Sb_0\left(x_y^v\right) \equiv x$

$Sb_{k+1}\left(x_y^v\right) \equiv Su\,[Sb_k(x_y^v)]\binom{k\,St\,v,x}{y}$

31. $Sb\left(x_y^v\right) \equiv Sb_{A(v,x)}\left(x_y^v\right)$[36]

$Sb\left(x_y^v\right)$ is the concept $Subst\,a\binom{v}{b}$, defined above.[37]

32. $x\,\mathrm{Imp}\,y \equiv [\mathrm{Neg}\,(x)]\,\mathrm{Dis}\,y$

$x\,\mathrm{Con}\,y \equiv \mathrm{Neg}\,\{[\mathrm{Neg}\,(x)]\,\mathrm{Dis}\,[\mathrm{Neg}\,(y)]\}$

$x\,\mathrm{Aeq}\,y \equiv (x\,\mathrm{Imp}\,y)\,\mathrm{Con}\,(y\,\mathrm{Imp}\,x)$

$v\,\mathrm{Ex}\,y \equiv \mathrm{Neg}\,\{v\,\mathrm{Gen}\,[\mathrm{Neg}\,(y)]\}$

33. $n\,Th\,x \equiv \varepsilon\,y\,\{y \leqq x^{(x^n)}\ \&\ (k)\big[k \leqq l(x) \rightarrow (k\,Gl\,x \leqq 13$

$\&\ k\,Gl\,y = k\,Gl\,x) \vee (k\,Gl\,x > 13$

$\&\ k\,Gl\,y = k\,Gl\,x\,.\,[1\,Pr\,(k\,Gl\,x)]^n)\big]\}$

$n\,Th\,x$ is the n-th *type-lift* of x (in the case when x and $n\,Th\,x$ are *formulae*).

To the axioms I, 1 to 3, there correspond three determinate numbers, which we denote by z_1, z_2, z_3, and we define:

34. $Z{-}A\,x(x) \equiv (x = z_1 \vee x = z_2 \vee x = z_3)$

---

[36] Where v is not a *variable* or x not a *formula*, then $Sb\left(x_y^v\right) = x$.

[37] Instead of $Sb\left[Sb\left(x_y^v\right)\frac{z}{w}\right]$ we write: $Sb\left(x\,_{y\,z}^{v\,w}\right)$ (and similarly for more than two *variables*).

35. $A_1 - A x (x) \equiv (Ey) [y \leqq x \ \& \ \text{Form}(y)$
$\& \ x = (y \operatorname{Dis} y) \operatorname{Imp} y]$

    x is a *formula* derived by substitution in the axiom-schema II, 1. Similarly $A_2 - Ax$, $A_3 - Ax$, $A_4 - Ax$ are defined in accordance with the axioms II, 2 to 4.

36. $A - A x (x) \equiv A_1 - A x (x) \vee A_2 - A x (x) \vee$
$A_3 - A x (x) \vee A_4 - A x (x)$

    x is a *formula* derived by substitution in an axiom of the sentential calculus.

37. $Q(z, y, v) \equiv (En, m, w) [n \leqq l(y) \ \& \ m \leqq l(z) \ \& \ w \leqq z$
$\& \ w = m \operatorname{Gl} x \ \& \ w \operatorname{Geb} n, y \ \& \ v \operatorname{Fr} n, y]$

    z contains no *variable bound* in y at a position where v is *free*.

38. $L_1 - A x (x) \equiv (Ev, y, z, n) \{v, y, z, n \leqq x \ \& \ n \operatorname{Var} v$
$\& \ \operatorname{Typ}_n(z) \ \& \ \operatorname{Form}(y) \ \& \ Q(z, y, v)$
$\& \ x = (v \operatorname{Gen} y) \operatorname{Imp} [Sb \left(\begin{smallmatrix} v \\ z \end{smallmatrix} \right)] \}$

    x is a *formula* derived from the axiom-schema III, 1 by substitution.

39. $L_2 - A x (x) \equiv (Ev, q, p) \{v, q, p \leqq x \ \& \ \operatorname{Var}(v)$
$\& \ \operatorname{Form}(p) \ \& \ v \operatorname{Fr} p \ \& \ \operatorname{Form}(q)$
$\& \ x = [v \operatorname{Gen}(p \operatorname{Dis} q)] \operatorname{Imp} [p \operatorname{Dis}(v \operatorname{Gen} q)] \}$

    x is a *formula* derived from the axiom-schema III, 2 by substitution.

40. $R - A x (x) \equiv (Eu, v, y, n) [u, v, y, n \leqq x \ \& \ n \operatorname{Var} v$
$\& \ (n + 1) \operatorname{Var} u \ \& \ u \operatorname{Fr} y \ \& \ \operatorname{Form}(y)$
$\& \ x = u \operatorname{Ex} \{v \operatorname{Gen} [[R(u) \ast E(R(v))] \operatorname{Aeq} y] \}]$

    x is a *formula* derived from the axiom-schema IV, 1 by substitution.

    To the axiom V, 1 there corresponds a determinate number z_4 and we define:

41. $M - A x (x) \equiv (En) [n \leqq x \ \& \ x = n \operatorname{Th} z_4]$

42. $A x(x) \equiv Z - A x(x) \vee A - A x(x) \vee L_1 - A x(x) \vee$
    $L_2 - A x(x) \vee R - A x(x) \vee M - A x(x)$
    x is an *axiom*.

43. $Fl(x\ y\ z) \equiv y = z \operatorname{Imp} x \vee (Ev)\ [v \leqq x\ \&\ \operatorname{Var}(v)$
    $\&\ x = v \operatorname{Gen} y]$
    x is an *immediate consequence* of y and z.

44. $Bw(x) \equiv (n)\ \{0 < n \leqq l(x) \rightarrow A x(n\,Gl\,x)$ [186]
    $\vee (Ep,q)\ [0 < p, q < n\ \&\ Fl(n\,Gl\,x, p\,Gl\,x, q\,Gl\,x)]\}$
    $\&\ l(x) > 0$
    x is a *proof-schema* (a finite series of *formulae*, of which
    each is either an *axiom* or an *immediate consequence*
    of two previous ones).

45. $x B y \equiv Bw(x)\ \&\ [l(x)]\,Gl\,x = y$
    x is a *proof* of the *formula* y.

46. $\operatorname{Bew}(x) = (Ey) y B x$
    x is a *provable formula*. [Bew (x) is the only one of the
    concepts 1–46 of which it cannot be asserted that it is
    recursive.]

The following proposition is an exact expression of a
fact which can be vaguely formulated in this way: every
recursive relation is definable in the system P (interpreted
as to content), regardless of what interpretation is given to
the formulae of P:

Proposition V: To every recursive relation $R\,(x_1 \ldots x_n)$
there corresponds an n-place *relation-sign* r (with the *free
variables*[38] $u_1, u_2, \ldots u_n$) such that for every n-tuple of
numbers $(x_1 \ldots x_n)$ the following hold:

---

[38] The *variables* $u_1 \ldots u_n$ could be arbitrarily allotted. There is
always, e.g., an r with the *free variables* 17, 19, 23 . . . etc., for which
(3) and (4) hold.

$$R(x_1 \ldots x_n) \rightarrow \mathrm{Bew}\left[Sb\left(r \begin{array}{ccc} u_1 & \ldots & u_n \\ Z(x_1) & \ldots & Z(x_n) \end{array} \right) \right] \tag{3}$$

$$R(x_1 \ldots x_n) \rightarrow \mathrm{Bew}\left[\mathrm{Neg}\, Sb\left(r \begin{array}{ccc} u_1 & \ldots & u_n \\ Z(x_1) & \ldots & Z(x_n) \end{array} \right) \right] \tag{4}$$

We content ourselves here with indicating the proof of this proposition in outline, since it offers no difficulties of principle and is somewhat involved.[39] We prove the proposition for all relations $R(x_1 \ldots x_n)$ of the form: $x_1 = \phi(x_2 \ldots x_n)$[40] (where ϕ is a recursive function) and apply mathematical induction on the degree of ϕ. For functions of the first degree (i.e. constants and the function $x+1$) the proposition is trivial. Let ϕ then be of degree m. It derives from functions of lower degree $\phi_1 \ldots \phi_k$ by the operations of substitution or recursive definition. Since, by the inductive assumption, everything is already proved for $\phi_1 \ldots \phi_k$, there exist corresponding *relation-signs* $r_1 \ldots r_k$ such that (3) and (4) hold. The processes of definition whereby ϕ is derived from $\phi_1 \ldots \phi_k$ (substitution and recursive definition) can all be formally mapped in the system P. If this is done, we obtain from $r_1 \ldots r_k$ a new *relation-sign* r[41], for which we can readily prove the validity of (3) and (4) by use of the inductive assumption. A *relation-sign* r, assigned in this fashion to a recursive relation,[42] will be called recursive.

[187]

We now come to the object of our exercises:

[39] Proposition V naturally is based on the fact that for any recursive relation R, it is decidable, for every n-tuple of numbers, **from the axioms of the system P**, whether the relation R holds or not.

[40] From this there follows immediately its validity for every recursive relation, since any such relation is equivalent to $0 = \phi(x_1 \ldots x_n)$, where ϕ is recursive.

[41] In the precise development of this proof, r is naturally defined, not by the roundabout route of indicating its content, but by its purely formal constitution.

[42] Which thus, as regards content, expresses the existence of this relation.

Let c be any class of *formulae*. We denote by Flg (c) (set of consequences of c) the smallest set of *formulae* which contains all the *formulae* of c and all *axioms*, and which is closed with respect to the relation "*immediate consequence of*". c is termed ω-consistent, if there is no *class-sign a* such that:

$$(n)\ [Sb\left(a \begin{smallmatrix} v \\ Z(n) \end{smallmatrix}\right) \varepsilon\ Flg(c)]\ \&\ [Neg(v\,Gen\,a)]\ \varepsilon\ Flg(c)$$

where v is the *free variable* of the *class-sign a*.

Every ω-consistent system is naturally also consistent. The converse, however, is not the case, as will be shown later.

The general result as to the existence of undecidable propositions reads:

Proposition VI: To every ω-consistent recursive class c of *formulae* there correspond recursive *class-signs r*, such that neither v Gen r nor Neg $(v$ Gen $r)$ belongs to Flg (c) (where v is the *free variable* of r).

Proof: Let c be any given recursive ω-consistent class of *formulae*. We define:

$$Bw_c(x) \equiv (n)\ [n \leqq l(x) \rightarrow Ax(n\,Gl\,x) \vee (n\,Gl\,x)\ \varepsilon\ c \vee$$
$$(Ep,q)\ \{0 < p, q < n\ \&\ Fl(n\,Gl\,x, p\,Gl\,x, q\,Gl\,x)\}]$$
$$\&\ l(x) > 0 \tag{5}$$

(cf. the analogous concept 44)

$$x\,B_c\,y \equiv Bw_c(x)\ \&\ [l(x)]\,Gl\,x = y \tag{6}$$

$$Bew_c(x) \equiv (Ey)\,y\,B_c\,x \tag{6.1}$$

(cf. the analogous concepts 45, 46)

The following clearly hold:

$$(x)\,[Bew_c(x) \sim x\,\varepsilon\,Flg(c)] \tag{7}$$

$$(x)\,[Bew(x) \rightarrow Bew_c(x)] \tag{8}$$

We now define the relation:

$$Q(x, y) \equiv \overline{x \, B_c \left[Sb \left(y \, {19 \atop Z(y)} \right) \right]}. \tag{8.1}$$

Since $x \, B_c \, y$ [according to (6), (5)] and $Sb \left(y \, {19 \atop Z(y)} \right)$ (according to definitions 17, 31) are recursive, so also is $Q(x, y)$. According to Proposition V and (8) there is therefore a *relation-sign q* (with the *free variables* 17, 19) such that

$$\overline{x \, B_c \left[Sb \left(y \, {19 \atop Z(y)} \right) \right]} \to \text{Bew}_c \left[Sb \left(q \, {17 \atop Z(x)} \, {19 \atop Z(y)} \right) \right] \tag{9}$$

$$x \, B_c \left[Sb \left(y \, {19 \atop Z(y)} \right) \right] \to \text{Bew}_c \left[\text{Neg } Sb \left(q \, {17 \atop Z(x)} \, {19 \atop Z(y)} \right) \right] \tag{10}$$

We put

$$p = 17 \, \text{Gen} \, q \tag{11}$$
$$(p \text{ is a } class\text{-}sign \text{ with the } free \ variable \ 19)$$

and

$$r = Sb \left(q \, {19 \atop Z(p)} \right) \tag{12}$$

(r is a recursive *class-sign* with the *free variable* 17).[43] Then

$$Sb \left(p \, {19 \atop Z(p)} \right) = Sb \left([17 \, \text{Gen} \, q] \, {19 \atop Z(p)} \right)$$

$$= 17 \, \text{Gen} \, Sb \left(q \, {19 \atop Z(p)} \right) \tag{13}$$

$$= 17 \, \text{Gen} \, r^{44}$$

---

[43] r is derived, in fact, from the recursive *relation-sign q* on replacement of a *variable* by a determinate number (p).

[44] The operations Gen and *Sb* are naturally always commutative, wherever they refer to different *variables*.

[because of (11) and (12)] and furthermore:

$$Sb\left(q \begin{array}{cc} 17 & 19 \\ Z(x) & Z(p) \end{array}\right) = Sb\left(r \begin{array}{c} 17 \\ Z(x) \end{array}\right) \qquad (14)$$

[according to (12)]. If now in (9) and (10) we substitute p for y, we find, in virtue of (13) and (14):

$$x \, B_c \, (17 \text{ Gen } r) \rightarrow \text{Bew}_c\left[Sb\left(r \begin{array}{c} 17 \\ Z(x) \end{array}\right)\right] \qquad (15)$$

$$x \, B_c \, (17 \text{ Gen } r) \rightarrow \text{Bew}_c\left[\text{Neg } Sb\left(r \begin{array}{c} 17 \\ Z(x) \end{array}\right)\right] \qquad (16)$$

Hence:

[189]

1. 17 Gen r is not *c-provable*.[45] For if that were so, there would (according to 6.1) be an n such that $n \, B_c \, (17 \text{ Gen } r)$. By (16) it would therefore be the case that:

$$\text{Bew}_c\left[\text{Neg } Sb\left(r \begin{array}{c} 17 \\ Z(n) \end{array}\right)\right]$$

while—on the other hand—from the *c-provability* of 17 Gen r there follows also that of $Sb\left(r \begin{array}{c} 17 \\ Z(n) \end{array}\right)$. c would therefore be inconsistent (and, *a fortiori*, ω-inconsistent).

2. Neg (17 Gen r) is not *c-provable*. Proof: As shown above, 17 Gen r is not *c-provable*, i.e. (according to 6.1) the following holds: $(n) \, n \, B_c \, (17 \text{ Gen } r)$. Whence it follows, by (15), that $(n) \, \text{Bew}_c\left[Sb\left(r \begin{array}{c} 17 \\ Z(n) \end{array}\right)\right]$, which together with $\text{Bew}_c \, [\text{Neg} \, (17 \text{ Gen } r)]$ would conflict with the ω-consistency of c.

17 Gen r is therefore undecidable in c, so that Proposition VI is proved.

---

[45] "x is *c-provable*" signifies: $x \, \varepsilon \, \text{Flg} \, (c)$, which, by (7), states the same as $\text{Bew}_c \, (x)$.

One can easily convince oneself that the above proof is constructive,[45a] i.e. that the following is demonstrated in an intuitionistically unobjectionable way: Given any recursively defined class c of *formulae*: If then a formal decision (in c) be given for the (effectively demonstrable) *propositional formula* 17 Gen r, we can effectively state:

1. A *proof* for Neg (17 Gen r).

2. For any given n, a *proof* for $Sb\left(r\begin{smallmatrix}17\\Z(n)\end{smallmatrix}\right)$, i.e. a formal decision of 17 Gen r would lead to the effective demonstrability of an ω-inconsistency.

We shall call a relation (class) of natural numbers $R(x_1 \ldots x_n)$ **calculable** [*entscheidungsdefinit*], if there is an n-place *relation-sign* r such that (3) and (4) hold (cf. Proposition V). In particular, therefore, by Proposition V, every recursive relation is calculable. Similarly, a *relation-sign* will be called **calculable**, if it be assigned in this manner to a calculable relation. It is, then, sufficient for the existence of undecidable propositions, to assume of the class c that it is ω-consistent and calculable. For the property of being calculable carries over from c to $x\ B_c\ y$ (cf. (5), (6)) and to $Q(x, y)$ (cf. 8.1), and only these are applied in the above proof. The undecidable proposition has in this case the form v Gen r, where r is a calculable *class-sign* (it is in fact enough that c should be calculable in the system extended by adding c).

[190]

If, instead of ω-consistency, mere consistency as such is assumed for c, then there follows, indeed, not the existence of an undecidable proposition, but rather the existence of a property (r) for which it is possible neither to provide a counter-example nor to prove that it holds for all numbers.

[45a] Since all existential assertions occurring in the proof are based on Proposition V, which, as can easily be seen, is intuitionistically unobjectionable.

For, in proving that 17 Gen r is not c-*provable*, only the consistency of c is employed (cf. p. 59) and from $\overline{\text{Bew}}_c$ (17 Gen r) it follows, according to (15), that for every number x, $Sb\left(r\, \dfrac{17}{Z(x)}\right)$ is c-*provable*, and hence that Neg $Sb\left(r\, \dfrac{17}{Z(x)}\right)$ is not c-*provable* for any number.

By adding Neg (17 Gen r) to c, we obtain a consistent but not ω-consistent *class of formulae* c'. c' is consistent, since otherwise 17 Gen r would be c-*provable*. c' is not however ω-consistent, since in virtue of $\overline{\text{Bew}}_c$ (17 Gen r) and (15) we have: $(x)\ \text{Bew}_c\ Sb\left(r\, \dfrac{17}{Z(x)}\right)$, and so *a fortiori*: $(x)\ \text{Bew}_{c'}\ Sb\left(r\, \dfrac{17}{Z(x)}\right)$, and on the other hand, naturally: $\text{Bew}_{c'}$ [Neg (17 Gen r)].[46]

A special case of Proposition VI is that in which the class c consists of a finite number of *formulae* (with or without those derived therefrom by *type-lift*). Every finite class α is naturally recursive. Let a be the largest number contained in α. Then in this case the following holds for c:

$$x \varepsilon c \sim (Em, n)\ [m \leqq x\ \&\ n \leqq a\ \&\ n \varepsilon \alpha\ \&\ x = m\, Th\, n]$$

c is therefore recursive. This allows one, for example, to conclude that even with the help of the axiom of choice (for all types), or the generalized continuum hypothesis, not all propositions are decidable, it being assumed that these hypotheses are ω-consistent.

In the proof of Proposition VI the only properties of the system P employed were the following:

[46] Thus the existence of consistent and not ω-consistent c's can naturally be proved only on the assumption that, in general, consistent c's do exist (i.e. that P is consistent).

1. The class of axioms and the rules of inference (i.e. the relation "immediate consequence of") are recursively definable (as soon as the basic signs are replaced in any fashion by natural numbers).

2. Every recursive relation is definable in the system P (in the sense of Proposition V).

Hence in every formal system that satisfies assumptions 1 and 2 and is ω-consistent, undecidable propositions exist of the form $(x)\,F(x)$, where F is a recursively defined property of natural numbers, and so too in every extension of such a system made by adding a recursively definable ω-consistent class of axioms. As can be easily confirmed, the systems which satisfy assumptions 1 and 2 include the Zermelo-Fraenkel and the v. Neumann axiom systems of set theory,[47] and also the axiom system of number theory which consists of the Peano axioms, the operation of recursive definition [according to schema (2)] and the logical rules.[48] Assumption 1 is in general satisfied by every system whose rules of inference are the usual ones and whose axioms (like those of P) are derived by substitution from a finite number of schemata.[48a]

[191]

[47] The proof of assumption 1 is here even simpler than that for the system P, since there is only one kind of basic variable (or two for J. v. Neumann).

[48] Cf. Problem III in D. Hilbert's lecture: 'Probleme der Grundlegung der Mathematik', *Math. Ann.* 102.

[48a] The true source of the incompleteness attaching to all formal systems of mathematics, is to be found—as will be shown in Part II of this essay—in the fact that the formation of ever higher types can be continued into the transfinite (cf. D. Hilbert, 'Über das Unendliche', *Math. Ann.* 95, p. 184), whereas in every formal system at most denumerably many types occur. It can be shown, that is, that the undecidable propositions here presented always become decidable by the adjunction of suitable higher types (e.g. of type ω for the system P). A similar result also holds for the axiom system of set theory.

3

From Proposition VI we now obtain further consequences and for this purpose give the following definition:

A relation (class) is called **arithmetical**, if it can be defined solely by means of the concepts $+$, $.$ [addition and multiplication, applied to natural numbers][49] and the logical constants $\vee$, $\overline{}$, (x), $=$, where (x) and $=$ are to relate only to natural numbers.[50] The concept of "arithmetical proposition" is defined in a corresponding way. In particular the relations "greater" and "congruent to a modulus" are arithmetical, since

$$x > y \sim \overline{(Ez)}\,[y = x + z]$$
$$x \equiv y\,(\mathrm{mod}\,n) \sim (Ez)\,[x = y + z\,.\,n \vee y = x + z\,.\,n]$$

We now have:

Proposition VII: Every recursive relation is arithmetical.

We prove this proposition in the form: Every relation of the form $x_0 = \phi\,(x_1 \ldots x_n)$, where ϕ is recursive, is arithmetical, and apply mathematical induction on the degree of ϕ. Let ϕ be of degree s ($s > 1$). Then either

1. $\phi\,(x_1 \ldots x_n) = \rho\,[\chi_1\,(x_1 \ldots x_n),$
   $\chi_2\,(x_1 \ldots x_n) \ldots \chi_m\,(x_1 \ldots x_n)]$[51]   [192]
   (where ρ and all the χ's have degrees smaller than s) or

2. $\phi\,(0, x_2 \ldots x_n) = \psi\,(x_2 \ldots x_n)$
   $\phi\,(k+1, x_2 \ldots x_n) = \mu\,[k, \phi\,(k, x_2 \ldots x_n), x_2 \ldots x_n]$
   (where ψ, μ are of lower degree than s).

[49] Here, and in what follows, zero is always included among the natural numbers.

[50] The definiens of such a concept must therefore be constructed solely by means of the signs stated, variables for natural numbers $x, y \ldots$ and the signs 0 and 1 (function and set variables must not occur). (Any other number-variable may naturally occur in the prefixes in place of x.)

[51] It is not of course necessary that all $x_1 \ldots x_n$ should actually occur in χ_i [cf. the example in footnote 27].

In the first case we have:

$$x_0 = \phi (x_1 \ldots x_n) \sim (Ey_1 \ldots y_m) [R (x_0\, y_1 \ldots y_m)$$
$$\& \; S_1 (y_1, x_1 \ldots x_n) \& \ldots \& S_m (y_m, x_1 \ldots x_n)],$$

where R and S_i are respectively the arithmetical relations which by the inductive assumption exist, equivalent to $x_0 = \rho (y_1 \ldots y_m)$ and $y = \chi_i (x_1 \ldots x_n)$. In this case, therefore, $x_0 = \phi (x_1 \ldots x_n)$ is arithmetical.

In the second case we apply the following procedure: The relation $x_0 = \phi (x_1 \ldots x_n)$ can be expressed with the help of the concept "series of numbers" (f)[52] as follows:

$$x_0 = \phi (x_1 \ldots x_n) \sim (Ef) \{f_0 = \psi (x_2 \ldots x_n)$$
$$\& \; (k) [k < x_1 \to f_{k+1} = \mu (k, f_k, x_2 \ldots x_n)]$$
$$\& \; x_0 = f_{x_1}\}$$

If $S (y, x_2 \ldots x_n)$ and $T (z, x_1 \ldots x_{n+1})$ are respectively the arithmetical relations—which by the inductive assumption exist—equivalent to

$$y = \psi (x_2 \ldots x_n) \text{ and } z = \mu (x_1 \ldots x_{n+1}),$$

the following then holds:

$$x_0 = \phi (x_1 \ldots x_n) \sim (Ef) \{S (f_0, x_2 \ldots x_n)$$
$$\& \; (k) [k < x_1 \to T (f_{k+1}, k, f_k, x_2 \ldots x_n)]$$
$$\& \; x_0 = f_{x_1}\} \tag{17}$$

We now replace the concept "series of numbers" by "pair of numbers", by assigning to the number pair n, d the number series $f^{(n,d)}$ $(f_k^{(n,d)} = [n]_{1+(k+1)d})$, where $[n]_p$ denotes the smallest non-negative residue of n modulo p.

---

[52] f signifies here a variable, whose domain of values consists of series of natural numbers. f_k denotes the $k+1$-th term of a series f (f_0 being the first).

We then have the following:

Lemma 1: If f is any series of natural numbers and k any natural number, then there exists a pair of natural numbers n, d, such that $f^{(n,d)}$ and f agree in the first k terms.

Proof: Let l be the largest of the numbers $k, f_0, f_1 \ldots f_{k-1}$. Let n be so determined that

$$n = f_i \, (\mathrm{mod} \, (1 + (i+1) \, l!)] \quad \text{for} \quad i = 0, 1 \ldots k-1$$

which is possible, since every two of the numbers $1 + (i+1)l!$ [193] $(i = 0, 1 \ldots k-1)$ are relatively prime. For a prime number contained in two of these numbers would also be contained in the difference $(i_1 - i_2) \, l!$ and therefore, because $|i_1 - i_2| < l$, in $l!$, which is impossible. The number pair n, $l!$ thus accomplishes what is required.

Since the relation $x = [n]_p$ is defined by $x \equiv n \, (\mathrm{mod} \, p)$ & $x < p$ and is therefore arithmetical, so also is the relation $P (x_0, x_1 \ldots x_n)$ defined as follows:

$$P (x_0 \ldots x_n) \equiv (En, d) \{ S ([n]_{d+1}, x_2 \ldots x_n)$$
$$\& (k) [k < x_1 \rightarrow T ([n]_{1+d(k+2)}, k, [n]_{1+d(k+1)},$$
$$x_2 \ldots x_n)] \& x_0 = [n]_{1+d(x_1+1)} \}$$

which, according to (17) and Lemma 1, is equivalent to $x_0 = \phi (x_1 \ldots x_n)$ (we are concerned with the series f in (17) only in its course up to the $x_1 + 1$-th term). Thereby Proposition VII is proved.

According to Proposition VII there corresponds to every problem of the form $(x) \, F (x)$ (F recursive) an equivalent arithmetical problem, and since the whole proof of Proposition VII can be formalized (for every specific F) within the system P, this equivalence is provable in P. Hence:

Proposition VIII: In every one of the formal systems[53] referred to in Proposition VI there are undecidable arithmetical propositions.

---

[53] These are the ω-consistent systems derived from P by addition of a recursively definable class of axioms.

The same holds (in virtue of the remarks at the end of Section 3) for the axiom system of set theory and its extensions by ω-consistent recursive classes of axioms.

We shall finally demonstrate the following result also:

Proposition IX: In all the formal systems referred to in Proposition VI[53] there are undecidable problems of the restricted predicate calculus[54] (i.e. formulae of the restricted predicate calculus for which neither universal validity nor the existence of a counter-example is provable).[55]

[194]   This is based on

Proposition X: Every problem of the form $(x) F(x)$ (F recursive) can be reduced to the question of the satisfiability of a formula of the restricted predicate calculus (i.e. for every recursive F one can give a formula of the restricted predicate calculus, the satisfiability of which is equivalent to the validity of $(x) F(x)$).

We regard the restricted predicate calculus (r.p.c.) as consisting of those formulae which are constructed out of the basic signs: $\overline{}$, $\vee$, (x), $=$; $x, y \ldots$ (individual variables) and $F(x)$, $G(x, y)$, $H(x, y, z) \ldots$ (property and relation variables)[56] where (x) and $=$ may relate only to individuals. To these signs we add yet a third kind of variables $\phi(x)$, $\psi(x\,y)$, $\chi(x\,y\,z)$ etc. which represent object functions; i.e.

---

[54] Cf. Hilbert-Ackermann, *Grundzüge der theoretischen Logik*. In the system P, formulae of the restricted predicate calculus are to be understood as those derived from the formulae of the restricted predicate calculus of PM on replacement of relations by classes of higher type, as indicated on p. 42.

[55] In my article 'Die Vollständigkeit der Axiome des logischen Funktionenkalküls', *Monatsh. f. Math. u. Phys.* XXXVII, 2, I have shown of every formula of the restricted predicate calculus that it is either demonstrable as universally valid or else that a counter-example exists; but in virtue of Proposition IX the existence of this counter-example is **not** always demonstrable (in the formal systems in question).

[56] D. Hilbert and W. Ackermann, in the work already cited, do not include the sign $=$ in the restricted predicate calculus. But for every formula in which the sign $=$ occurs, there exists a formula without this sign, which is satisfiable simultaneously with the original one (cf. the article cited in footnote 55).

$\phi\,(x)$, $\psi\,(x\,y)$, etc. denote one-valued functions whose arguments and values are individuals.[57] A formula which, besides the first mentioned signs of the r.p.c., also contains variables of the third kind, will be called a formula in the wider sense (i.w.s.).[58] The concepts of "satisfiable" and "universally valid" transfer immediately to formulae i.w.s. and we have the proposition that for every formula i.w.s. A we can give an ordinary formula of the r.p.c. B such that the satisfiability of A is equivalent to that of B. We obtain B from A, by replacing the variables of the third kind $\phi\,(x)$, $\psi\,(x\,y)$ ... appearing in A by expressions of the form $(\imath\,z)\,F\,(z\,x)$, $(\imath\,z)\,G\,(z,\,x\,y)$ ..., by eliminating the "descriptive" functions on the lines of PM I $*$ 14, and by logically multiplying[59] the resultant formula by an expression, which states that all the F, G ... substituted for the ϕ, ψ ... are strictly one-valued with respect to the first empty place.

We now show, that for every problem of the form $(x)\,F\,(x)$ (F recursive) there is an equivalent concerning the satisfiability of a formula i.w.s., from which Proposition X follows in accordance with what has just been said.

Since F is recursive, there is a recursive function $\Phi\,(x)$ such that $F\,(x) \sim [\Phi\,(x) = 0]$, and for Φ there is a series of functions Φ_1, Φ_2 ... Φ_n, such that $\Phi_n = \Phi$, $\Phi_1\,(x) = x+1$ and for every Φ_k $(1 < k \leqq n)$ either

$$
\begin{aligned}
&1. \ (x_2 \ldots x_m)\,[\Phi_k\,(0,\,x_2 \ldots x_m) = \Phi_p\,(x_2 \ldots x_m)] \\
&\quad (x,\,x_2 \ldots x_m)\,\{\Phi_k\,[\Phi_1\,(x),\,x_2 \ldots x_m] \\
&\quad\quad = \Phi_q\,[x,\,\Phi_k\,(x,\,x_2 \ldots x_m),\,x_2 \ldots x_m]\} \qquad (18) \\
&\quad\quad\quad p,\,q < k
\end{aligned}
$$

[57] And of course the domain of the definition must always be the **whole** domain of individuals.

[58] Variables of the third kind may therefore occur at all empty places instead of individual variables, e.g. $y = \phi\,(x)$, $F\,(x,\,\phi\,(y))$, $G\,[\psi\,(x,\,\phi\,(y)),\,x]$ etc.

[59] I.e. forming the conjunction.

or

2. $(x_1 \ldots x_m) [\Phi_k(x_1 \ldots x_m) = \Phi_r(\Phi_{i_1}(\mathfrak{x}_1) \ldots \phi_{i_s}(\mathfrak{x}_s))]^{60}$ (19)
$$r < k, \; i_v < k \quad (\text{for } v = 1, 2 \ldots s)$$

or

3. $(x_1 \ldots x_m) [\Phi_k(x_1 \ldots x_m) = \Phi_1(\Phi_1 \ldots \Phi_1(0))]$ (20)

In addition, we form the propositions:

$$(x) \overline{\Phi_1(x) = 0} \; \& \; (x\,y) [\Phi_1(x) = \Phi_1(y) \to x = y] \quad (21)$$
$$(x) [\Phi_n(x) = 0] \quad (22)$$

In all the formulae (18), (19), (20) (for $k = 2, 3, \ldots n$) and in (21), (22), we now replace the functions Φ_i by the function variable ϕ_i, the number 0 by an otherwise absent individual variable x_0 and form the conjunction C of all the formulae so obtained.

The formula $(E\,x_0)\,C$ then has the required property, i.e.

1. If $(x) [\Phi(x) = 0]$ is the case, then $(E\,x_0)\,C$ is satisfiable, since when the functions $\Phi_1, \Phi_2, \ldots \Phi_n$ are substituted for $\phi_1, \phi_2, \ldots \phi_n$ in $(E\,x_0)\,C$ they obviously yield a correct proposition.

2. If $(E\,x_0)\,C$ is satisfiable, then $(x) [\Phi(x) = 0]$ is the case.

Proof: Let $\Psi_1, \Psi_2 \ldots \Psi_n$ be the functions presumed to exist, which yield a correct proposition when substituted for $\phi_1, \phi_2 \ldots \phi_n$ in $(E\,x_0)\,C$. Let its domain of individuals be I. In view of the correctness of $(E\,x_0)\,C$ for all functions Ψ_i, there is an individual a (in I) such that all the formulae (18) to (22) transform into correct propositions (18') to (22') on replacement of the Φ_i by Ψ_i and of 0 by a. We now form the smallest sub-class of I, which contains a and is closed with respect to the operation $\Psi_1(x)$. This sub-class (I') has the property that every one of the functions

---

[60] $\mathfrak{x}_i \, (i = 1 \ldots s)$ represents any complex of the variables $x_1, x_2 \ldots x_m$, e.g. $x_1 \, x_3 \, x_2$.

Ψ_i, when applied to elements of I', again yields elements of I'. For this holds of Ψ_1 in virtue of the definition of I'; and by reason of (18'), (19'), (20') this property carries over from Ψ_i of lower index to those of higher. The functions derived from Ψ_i by restriction to the domain of individuals I', we shall call Ψ_i'. For these functions also the formulae (18) to (22) all hold (on replacement of 0 by a and Φ_i by Ψ_i').

Owing to the correctness of (21) for Ψ_1' and a, we can map the individuals of I' in one-to-one correspondence on the natural numbers, and this in such a manner that a transforms into 0 and the function Ψ_1' into the successor function Φ_1. But, by this mapping, all the functions Ψ_i' transform into the functions Φ_i, and owing to the correctness of (22) for Ψ'_n and a, we get $(x) [\Phi_n(x) = 0]$ or $(x) [\Phi(x) = 0]$, which was to be proved.[61]   [196]

Since the considerations leading to Proposition X (for every specific F) can also be restated within the system P, the equivalence between a proposition of the form $(x) F(x)$ (F recursive) and the satisfiability of the corresponding formula of the r.p.c. is therefore provable in P, and hence the undecidability of the one follows from that of the other, whereby Proposition IX is proved.[62]

4

From the conclusions of Section 2 there follows a remarkable result with regard to a consistency proof of the system

---

[61] From Proposition X it follows, for example, that the Fermat and Goldbach problems would be soluble, if one had solved the decision problem for the r.p.c.

[62] Proposition IX naturally holds also for the axiom system of set theory and its extensions by recursively definable ω-consistent classes of axioms, since in these systems also there certainly exist undecidable theorems of the form $(x) F(x)$ (F recursive).

P (and its extensions), which is expressed in the following proposition:

Proposition XI: If c be a given recursive, consistent class[63] of *formulae*, then the *propositional formula* which states that c is consistent is not *c-provable*; in particular, the consistency of P is unprovable in P,[64] it being assumed that P is consistent (if not, of course, every statement is provable).

The proof (sketched in outline) is as follows: Let c be any given recursive class of *formulae*, selected once and for all for purposes of the following argument (in the simplest case it may be the null class). For proof of the fact that 17 Gen r is not *c-provable*,[65] only the consistency of c was made use of, as appears from 1, page 59; i.e.

$$\text{Wid}(c) \to \overline{\text{Bew}_c}(17 \text{ Gen } r) \qquad (23)$$

i.e. by (6.1):

$$\text{Wid}(c) \to (x)\, \overline{x\, B_c\, (17 \text{ Gen } r)}$$

By (13), $17 \text{ Gen } r = Sb\left(p \begin{array}{c} 19 \\ Z(p) \end{array}\right)$ and hence:

$$\text{Wid}(c) \to (x)\, \overline{x\, B_c\, Sb\left(p \begin{array}{c} 19 \\ Z(p) \end{array}\right)}$$

[197]

i.e. by (8.1):

$$\text{Wid}(c) \to (x)\, Q(x, p) \qquad (24)$$

We now establish the following: All the concepts defined (or assertions proved) in Sections 2[66] and 4 are also expressible (or provable) in P. For we have employed through-

---

[63] c is consistent (abbreviated as $\text{Wid}(c)$) is defined as follows: $\text{Wid}(c) = (E\,x)\,[\text{Form}(x)\,\&\,\overline{\text{Bew}_c}(x)]$.

[64] This follows if c is replaced by the null class of *formulae*.

[65] r naturally depends on c (just as p does).

[66] From the definition of "recursive" on p. 46 up to the proof of Proposition VI inclusive.

out only the normal methods of definition and proof accepted in classical mathematics, as formalized in the system P. In particular c (like any recursive class) is definable in P. Let w be the *propositional formula* expressing Wid (c) in P. The relation $Q(x, y)$ is expressed, in accordance with (8.1), (9) and (10), by the *relation-sign* q, and $Q(x, p)$, therefore, by $r \left[\text{ since by (12) } r = Sb\left(q \begin{array}{c} 19 \\ Z(p) \end{array} \right) \right]$ and the proposition $(x) \, Q(x, p)$ by 17 Gen r.

In virtue of (24) w Imp (17 Gen r) is therefore *provable* in P[67] (and *a fortiori c-provable*). Now if w were *c-provable*, 17 Gen r would also be *c-provable* and hence it would follow, by (23), that c is not consistent.

It may be noted that this proof is also constructive, i.e. it permits, if a *proof* from c is produced for w, the effective derivation from c of a contradiction. The whole proof of Proposition XI can also be carried over word for word to the axiom-system of set theory M, and to that of classical mathematics A,[68] and here too it yields the result that there is no consistency proof for M or for A which could be formalized in M or A respectively, it being assumed that M and A are consistent. It must be expressly noted that Proposition XI (and the corresponding results for M and A) represent no contradiction of the formalistic standpoint of Hilbert. For this standpoint presupposes only the existence of a consistency proof effected by finite means, and there might conceivably be finite proofs which **cannot** be stated in P (or in M or in A).

Since, for every consistent class c, w is not *c-provable*, there will always be propositions which are undecidable

[67] That the correctness of w Imp (17 Gen r) can be concluded from (23), is simply based on the fact that—as was remarked at the outset—the undecidable proposition 17 Gen r asserts its own unprovability.

[68] Cf. J. v. Neumann, 'Zur Hilbertschen Beweistheorie', *Math. Zeitschr.* 26, 1927.

[198]

(from c), namely w, so long as Neg (w) is not *c-provable*; in other words, one can replace the assumption of ω-consistency in Proposition VI by the following: The statement "c is inconsistent" is not c-provable. (Note that there are consistent c's for which this statement is c-provable.)

Throughout this work we have virtually confined ourselves to the system P, and have merely indicated the applications to other systems. The results will be stated and proved in fuller generality in a forthcoming sequel. There too, the mere outline proof we have given of Proposition XI will be presented in detail.

(Received : 17 . xi . 1930.)